WHAT'S NEXT ?

Volume 1

Using Patterns
to
Solve Problems

Wilbert Reimer
Fresno Pacific University

Elaine Reimer Paré, Editor

Reneé Mason, Illustrator
Brenda Wood, Cover Design
Roxanne Williams, Desktop Publisher
Leticia Rivera, Desktop Publisher

AIMS Education Foundation
Fresno, California

This book contains materials developed by the AIMS Education Foundation. **AIMS** (**A**ctivities **I**ntegrating **M**athematics and **S**cience) began in 1981 with a grant from the National Science Foundation. The non-profit AIMS Education Foundation publishes hands-on instructional materials (books and the quarterly magazine) that integrate curricular disciplines such as mathematics, science, language arts, and social studies. The Foundation sponsors a national program of professional development through which educators may gain both an understanding of the AIMS philosophy and expertise in teaching by integrated, hands-on methods.

ISBN: **978-1-881431-54-1**

Printed in the United
States of America

TABLE OF CONTENTS

INTRODUCTION

APPENDIX

FIGURING OUT *WHAT'S NEXT ?*

Searching for patterns might well be the most important experience the student will have in mathematics.

Using patterns to make discoveries leads students to a greater understanding of both mathematics and the world. *What's Next?* implements a pattern discovery approach to problem solving. Difficult problems become simpler when order and relationships are uncovered. With such an approach, "students have opportunities to generalize and describe patterns and functions in many ways and to explore the relationships among them" (National Council of Teachers of Mathematics,1989). This series should inspire students to learn, understand, and continually ask "What's next?"

Each volume of *What's Next?* contains life-related problems that address concepts from every area of mathematics. They are adaptable to a wide range of student abilities and may be explored in any sequence. Students gain valuable experience by seeing and identifying patterns. Teachers may want to begin with those activities that invite students to draw the next figure or guess the next number in a sequence. These will build pattern recognition confidence.

Many activities ask students to construct a table and continue a pattern. Once data are collected and recorded in a table, the pattern becomes easier to recognize. Students use the pattern they discover to complete the table and solve the problem. Often students are challenged to find a general formula for the problem. This step "builds readiness for a generalized view of mathematics" (National Council of Teachers of Mathematics, 1989).

Students should be familiar with the dominant patterns in mathematics such as those in square numbers, triangular numbers, and Pascal's Triangle. These and other helpful tools for understanding the role of patterns in problem solving are provided in the *Appendix* and the *Solutions*.

The Wishful Thinking Method

Dr. George Pólya, renowned author and teacher of problem solving, often advised, "If you have a difficult problem, wish for an easier one!" This "wish" is actually the first step of a powerful method for making discoveries. The wishful thinking method is encouraged throughout the activities in *What's Next*? Students are often asked to consider a simpler problem first. Using patterns makes the original problem surprisingly easy!

Vertical Solutions:

When working on a problem that uses a table, look for a vertical pattern in the second column. Examine the differences of the numbers in the second column. If this does not reveal a pattern, look at the differences of the differences, etc. This process works amazingly well. Remember, the *differences* are the key!

For many students, finding the vertical pattern in the second column will provide an appropriate challenge. The activities serve as excellent problem-solving experiences without moving beyond this level.

Horizontal Solutions:

Students ready to advance to the next level of problem solving may examine the horizontal relationship between the numbers in the columns. They should ask, "What must be done to the number in the first column to obtain the number in the second column?" Once the horizontal relationship has been discovered, it may be expressed in mathematical language as a generalizing formula.

A sample problem will illustrate how the wishful thinking method works. More hints for finding patterns appear in the *Solutions.*

Sample Problem:

Find the sum of the first 50 even numbers.

$$2 + 4 + 6 + 8 + 10 + \ldots = ?$$

A first reaction to this problem might be: "I wish I had an easier problem!"

Simplify the problem and record your discoveries in a table.

Number of Terms	Sum
1	2
2	6
3	12
4	20
5	30
n	n(n+1)

Find the sums of the series with one term, two terms, three terms, and so on. Continue adding and recording your results in the table.

Looking at a number of simpler cases is the first step of the wishful thinking problem-solving approach.

Vertical solution for sample problem:

Examine the numbers in the second column (2, 6, 12, 20, and 30). Notice that their differences are 4, 6, 8, and 10. A vertical pattern has been discovered! This discovery makes it possible to continue the vertical pattern in the second column indefinitely.

Horizontal solution for sample problem:

The horizontal solution challenges the student to discover what must be done to the number in the first column to get the number in the second column. In this case, the student must multiply the number in the first column by a number one larger. In mathematical language, the horizontal solution would be written as $n(n+1)$, where n represents any number in the first column.

What's Next for you?

Complete solutions to every activity are provided in the back of this book, along with suggestions and ideas for making the experience of pattern recognition valuable and practical. Don't be surprised if students discover patterns beyond the ones identified in the *Solutions.* One of the joys of mathematics—especially of problem solving—is that there is usually more than one way to solve a problem. Use these activities to stimulate and encourage student creativity. What's next? Hopefully, the satisfaction of understanding and the joy of discovery!

National Council of Teachers of Mathematics. *Curriculum and Evaluation Standards for School Mathematics.* Reston, VA.: NCTM, 1989.

PRACTICALLY PREDICTABLE !

Discover the pattern and continue each series.

1. 30, 25, 20 , ______ , ______ , ______
2. January, March, May , ______ , ______ , ______
3. 6, 4, 2, 0, -2 , ______ , ______ , ______ , ______
4. 1/2 , 1/4 , 1/8 , ______ , ______ , ______
5. z, y, w, t , ______ , ______ , ______
6. Wednesday, Saturday, Tuesday, Friday , ______ , ______
7. 7700, 6600, 5500 , ______ , ______ , ______
8. .125 , .25 , .375 , .5 , ______ , ______ , ______
9. (Uncle, nephew) , (Aunt, ______)
10. 1 x 1/2 , 2 x 1/4 , 3 x 1/6 , __________ , __________
11. .1 , .1 , .2 , .3 , .5 , ______ , ______ , ______ , ______
12. ______
13. ______
14. ______

IN AND OUT

Discover the "ins and outs" of these patterns by completing the tables.

In	Out
math	m
zebra	z
house	h
pick	p
school	
mouse	
nose	

In	Out
guessing	9
involves	9
taking	7
a	2
risk	5
but	4
it	3
is	
often	
a	
good	
strategy	

In	Out
problem	13
solving	7
in	14
math	8
can	14
be	5
much	8
fun	14
do	
you	
think	
so	

In	Out
bath	b
cheetah	f
love	p
line	j
stem	f
elephant	f
sit	j
pin	
pickle	
today	
think	

In	Out
3	7
10	21
5	11
0	1
50	101
4	9
15	
6	
20	
	61
	201

In	Out
2	2
145	10
31	4
10	1
8	8
182	11
0	0
20	
3	
481	
16	

ROLLER COASTER SUMS

Like a roller coaster, each of the following series climbs up, then plunges to its starting position. Find the sums for the "roller coaster" series.

Do you notice a pattern? Use your discovery to sum the rest of the series.

$\boxed{1}$ = ________

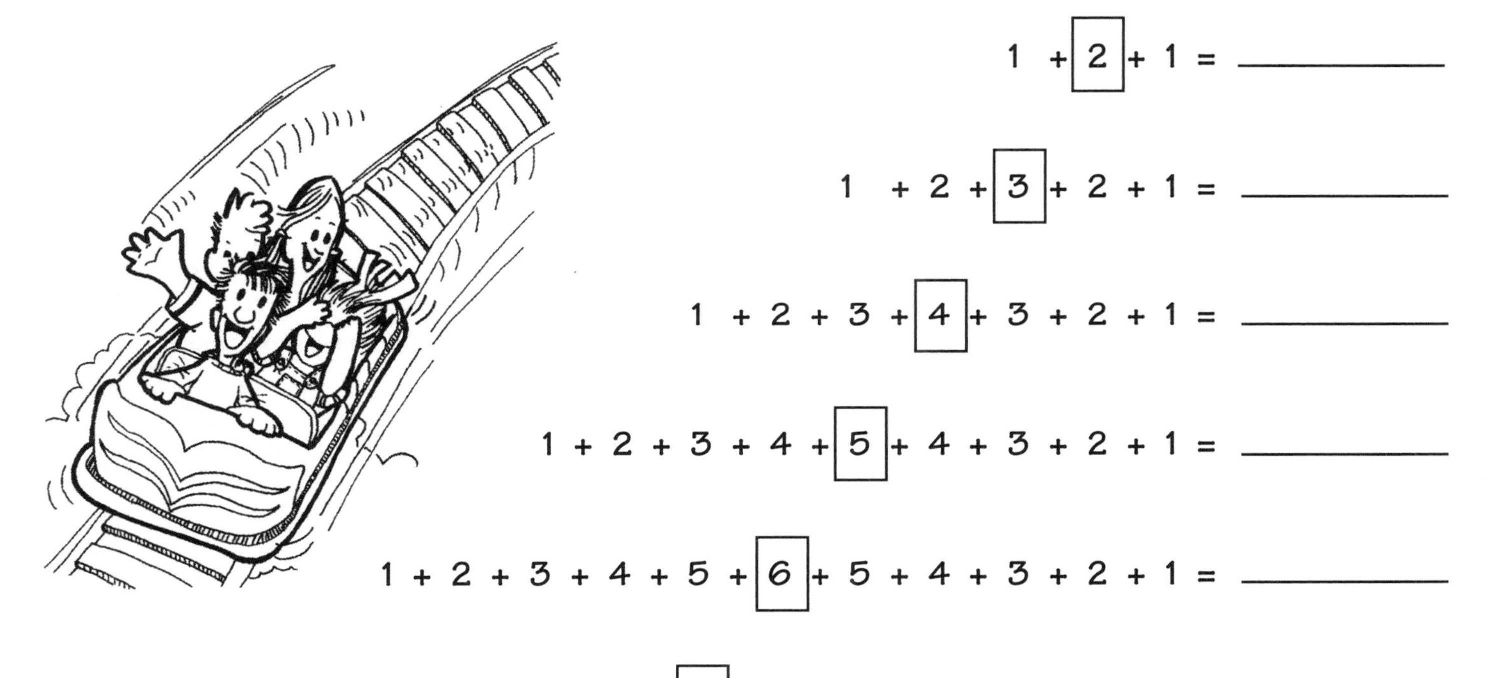

$1 + \boxed{2} + 1$ = ________

$1 + 2 + \boxed{3} + 2 + 1$ = ________

$1 + 2 + 3 + \boxed{4} + 3 + 2 + 1$ = ________

$1 + 2 + 3 + 4 + \boxed{5} + 4 + 3 + 2 + 1$ = ________

$1 + 2 + 3 + 4 + 5 + \boxed{6} + 5 + 4 + 3 + 2 + 1$ = ________

$1 + 2 + 3 + 4 + 5 + 6 + \boxed{7} + 6 + 5 + 4 + 3 + 2 + 1$ = ________

$1 + 2 + 3 + 4 + 5 + 6 + 7 + \boxed{8} + 7 + 6 + 5 + 4 + 3 + 2 + 1$ = ________

EXTRA CHALLENGE:
If the highest number is 20, what is the sum of the "roller coaster" series? What is the sum of the series if the highest number is n?

SQUARE, OBLONG, AND TRIANGULAR NUMBERS

Square numbers are numbers that can be represented by dots in a square array. The first four square numbers are pictured below.

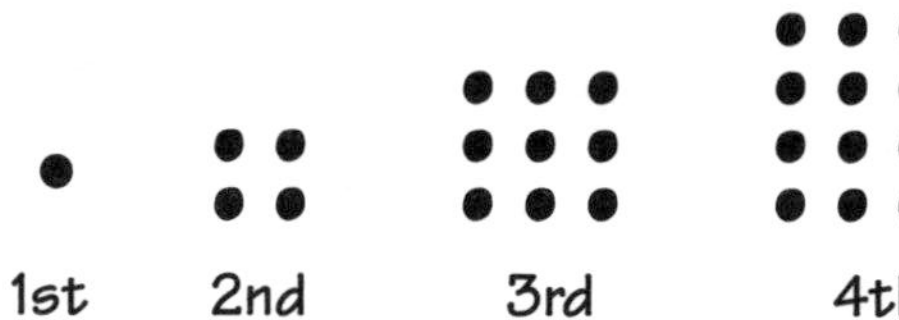

Square Number	Number of Dots
1st	1
2nd	4
3rd	______
4th	______
5th	______
6th	______
50th	______
nth	______

Complete the table to find the number of dots in the nth square number.

Oblong numbers are numbers that can be represented by dots in a rectangle having one dimension one unit longer than the other. The first four oblong numbers are pictured below.

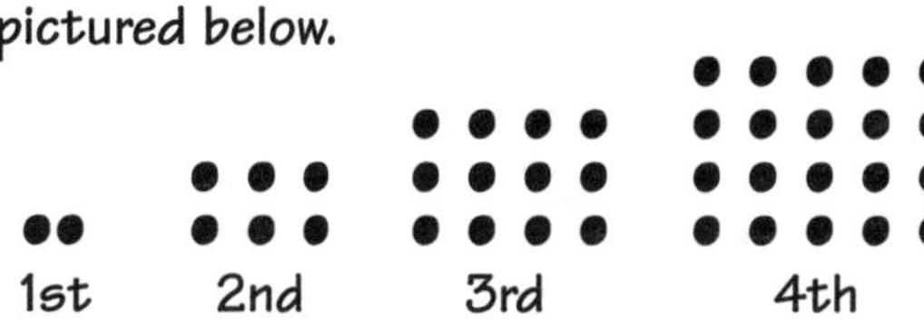

Oblong Number	Number of Dots
1st	2
2nd	6
3rd	______
4th	______
5th	______
6th	______
50th	______
nth	______

Complete the table to find the number of dots in the nth oblong number.

Triangular numbers are numbers that can be represented by dots in a triangular array. The first four triangular numbers are pictured below.

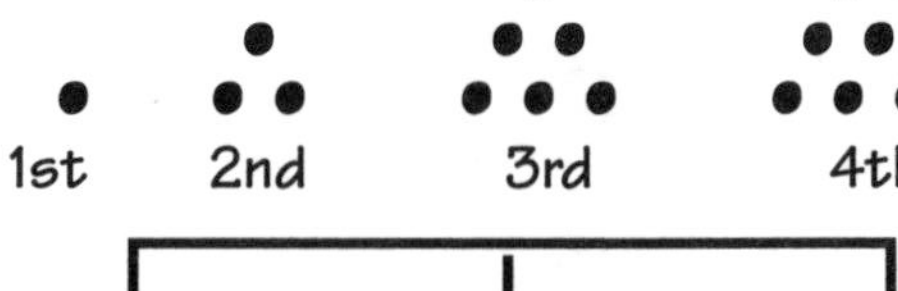

Triangular Number	Number of Dots
1st	1
2nd	3
3rd	______
4th	______
5th	______
6th	______
50th	______
nth	______

Complete the table to find the number of dots in the nth triangular number.

TABLE IT !

Complete the following tables.

0	8
1	11
2	14
3	17
4	
5	
50	
n	

0	2
1	7
2	12
3	17
4	
5	
50	
n	

1	4
2	7
3	12
4	19
5	
6	
50	
n	

0	4
1	9
2	16
3	25
4	
5	
50	
n	

1	0
2	8
3	16
4	24
5	
6	
50	
n	

1	1
2	8
3	27
4	64
5	
6	
50	
n	

Make up some tables for your friends to complete.

FIGURATIVELY SPEAKING

Figurate number families—numbers which can be represented by dots in specific shapes—may be found by summing sequences.

For example, here are the first four pentagonal numbers:

1st 2nd 3rd 4th

Study the pattern in this table, which shows how triangular, square, and pentagonal numbers may be generated.

Number Family	Triangular	Square	Pentagonal
1st	1=1	1=1	1=1
2nd	1+2=3	1+3=4	1+4=5
3rd	1+2+3=6	1+3+5=9	1+4+7=12
4th	1+2+3+4=10	1+3+5+7=16	1+4+7+10=22
5th	1+2+3+4+5=15	1+3+5+7+9=25	1+4+7+10+13=35

Use these patterns to predict the first five hexagonal, heptagonal and octagonal numbers.

Number Family	Hexagonal	Heptagonal	Octagonal
1st			
2nd			
3rd			
4th			
5th			

CHINESE CHECKERS

A Chinese Checkers board has six starting pens (triangular regions making the points of the star). Each pen has four rows of holes which hold ten marbles.

Suppose a board were constructed that had seven rows of holes in each starting pen.

What would be the total number of holes on this expanded board?

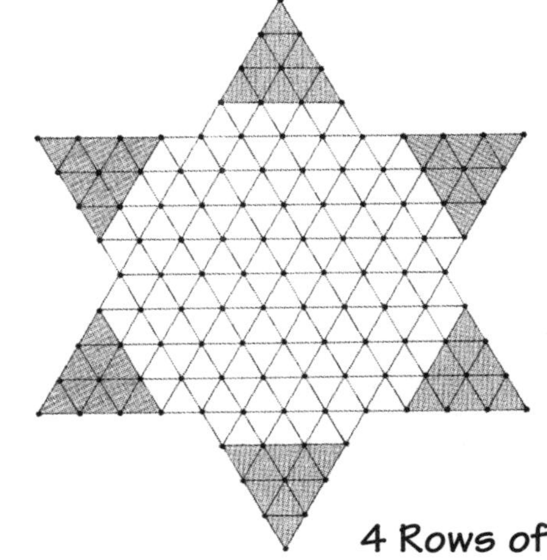

4 Rows of Holes

Look at simpler versions of the problem to observe a pattern. As soon as you see the pattern, complete the table.

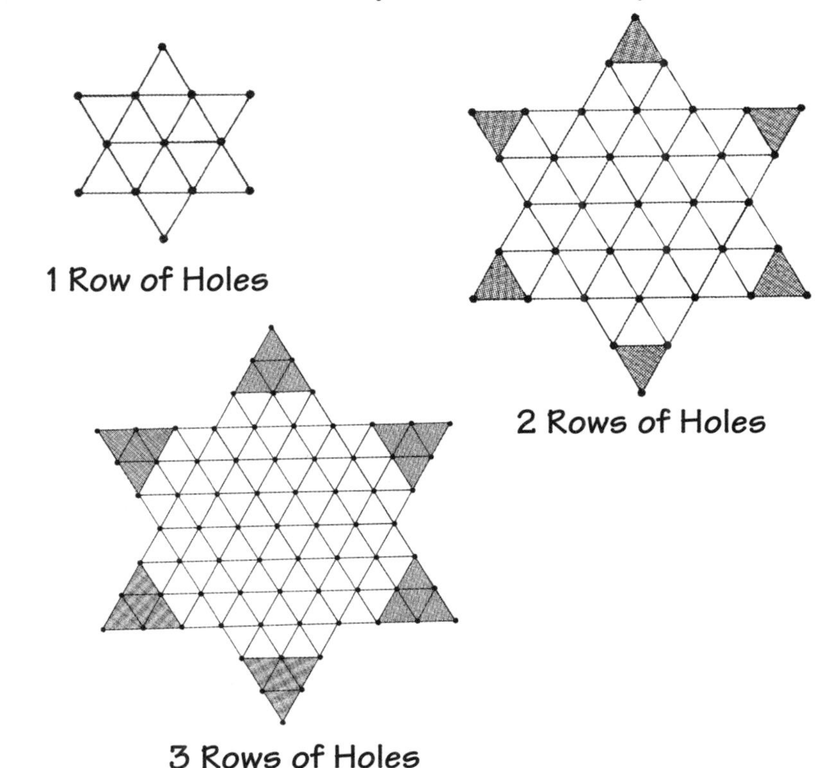

1 Row of Holes

2 Rows of Holes

3 Rows of Holes

Number of Rows of Holes in Starting Pen	Total Number of Holes
1	13
2	
3	
4	
5	
6	
7	

EXTRA CHALLENGE:
What is the total number of holes when each starting pen has n rows of holes?

CALENDAR CROSSING

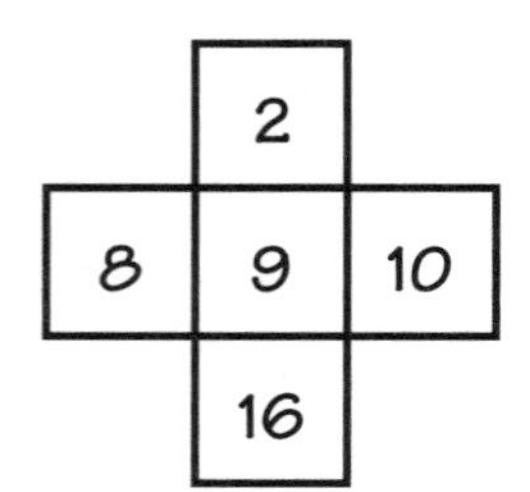

	10	
16	17	18
	24	

Find more crosses like these on the calendars.

Discover the relationship of the center number of each cross to the sum of all five numbers in the cross. Complete the table.

Center Number	Sum of the Five Numbers
9	45
n	

Sun	Mon	Tue	Wed	Thur	Fri	Sat
		1	2	3	4	5
6	7	8	9	10	11	12
13	14	15	16	17	18	19
20	21	22	23	24	25	26
27	28	29	30	31		

Sun	Mon	Tue	Wed	Thur	Fri	Sat
	1	2	3	4	5	6
7	8	9	10	11	12	13
14	15	16	17	18	19	20
21	22	23	24	25	26	27
28	29	30	31			

Sun	Mon	Tue	Wed	Thur	Fri	Sat
			1	2	3	4
5	6	7	8	9	10	11
12	13	14	15	16	17	18
19	20	21	22	23	24	25
26	27	28	29	30	31	

Sun	Mon	Tue	Wed	Thur	Fri	Sat
					1	2
3	4	5	6	7	8	9
10	11	12	13	14	15	16
17	18	19	20	21	22	23
24	25	26	27	28	29	30
31						

Look at a calendar for any month of any year. Is your discovery about calendar crosses always true?

BUSY INTERSECTIONS

When ten points are placed on a circle and all possible line segments have been drawn, what is the maximum number of intersections?

To solve this problem, count the number of intersections in circles with four, five, six, and seven points. Find this pattern in Pascal's triangle and predict the solution.

Number of Points	Number of Intersections
4	1
5	5
6	
7	
8	
9	
10	

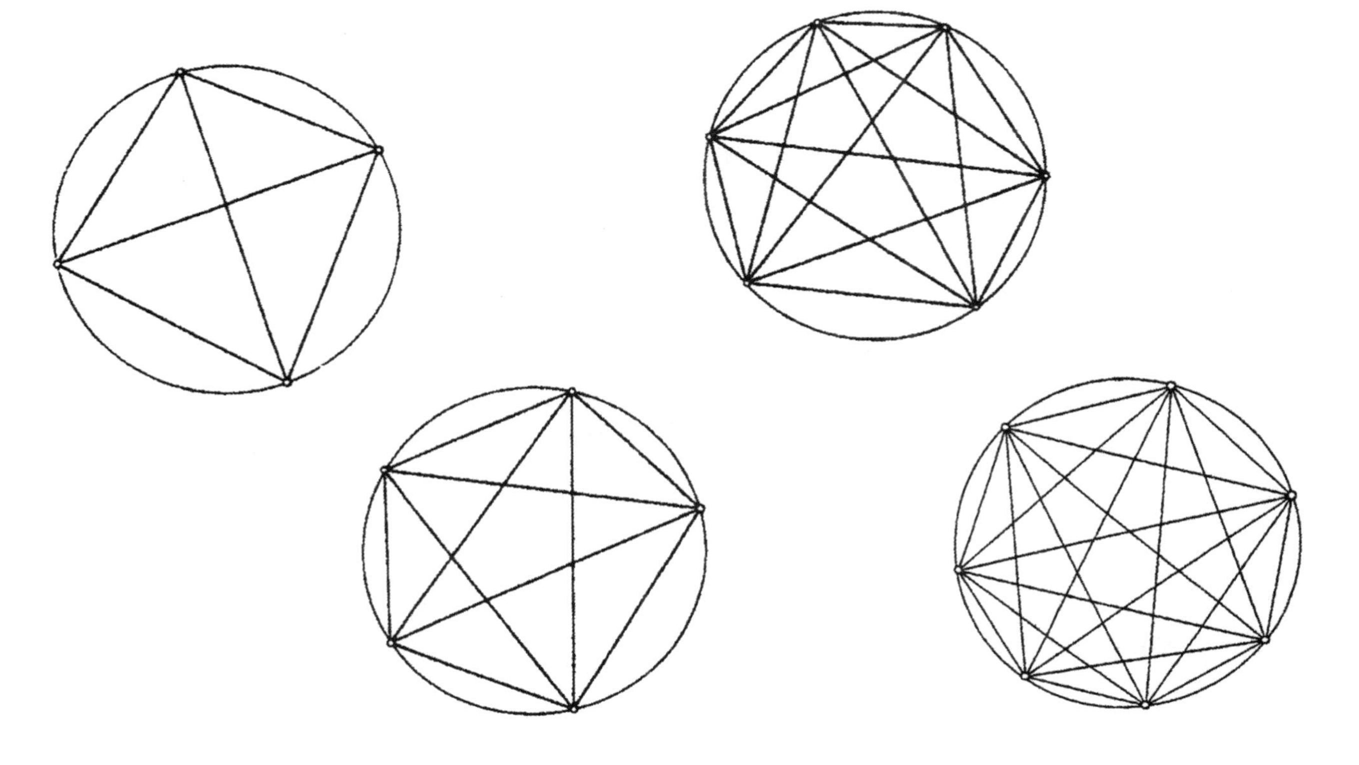

THE FOLD THAT FOOLS

Imagine folding one large piece of paper in half. If you fold your result in half again and continue this process until you have folded it 50 times, how many layers of paper will you have?

Complete the table to discover the amazing result!

Number of Folds	Number of Layers
0	1
1	2
2	
3	
4	
5	
6	
50	
n	

EXTRA CHALLENGE:
If a piece of paper is 1/1000 of an inch thick and folded in half 50 times, how tall would the folded stack be?

TRIPLE YOUR FUN !

Discover the patterns in the triples and fill in the missing numbers.

Triple Pattern #1
(1, 4, 6)
(5, 8, 10)
(0, 3, 5)
(4, ___ , ___)
(12, ___ , ___)
(___ , 8, ___)
(___ , ___ , 28)
(___ , 60, ___)

Triple Pattern #2
(2, 4, 8)
(1, 1, 1)
(4, 16, 64)
(3, 9, 27)
(5, ___ , ___)
(0, ___ , ___)
(___ , 64, ___)
(7, ___ , ___)

Triple Pattern #3
(1, 1, 1)
(2, 3, 4)
(3, 6, 9)
(4, 10, 16)
(5, ___ , ___)
(6, ___ , ___)
(___ , ___ , 100)
(___ , ___ , 121)

Triple Pattern #4
(8, 2, 7)
(16, 4, 9)
(28, 7, 12)
(40, 10, 15)
(32, ___ , 13)
(12 , ___ , ___)
(___ , 5, ___)
(44 , ___ , ___)

Triple Pattern #5
(15, 5, 10)
(27, 9, 18)
(12, 4, 8)
(30, 10, 20)
(9, ___ , 6)
(___ , 7, ___)
(18, ___ , ___)
(___ , ___ , 24)

Make up some of your own triples. Can your friends discover the patterns?

THAT FIGURES !

If these patterns are continued, how many dots will be in the 100th figure of each sequence?

1.

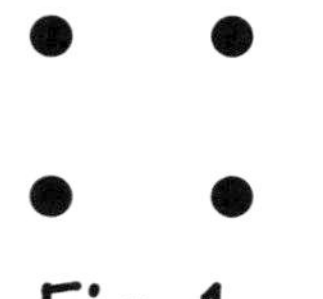
Fig. 1

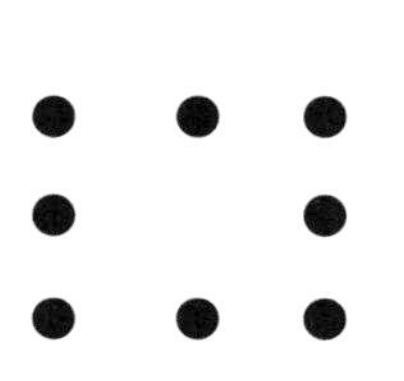
Fig.2

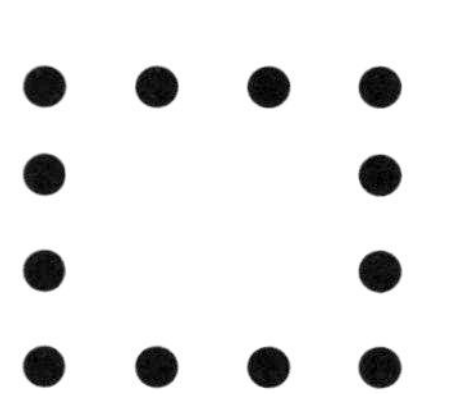
Fig. 3

2.

Fig. 1

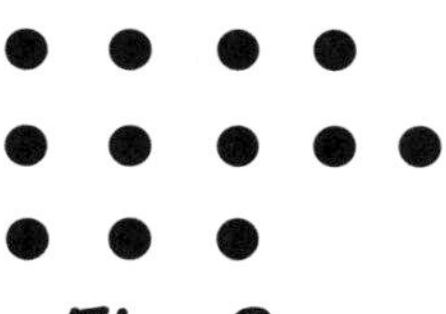
Fig. 2

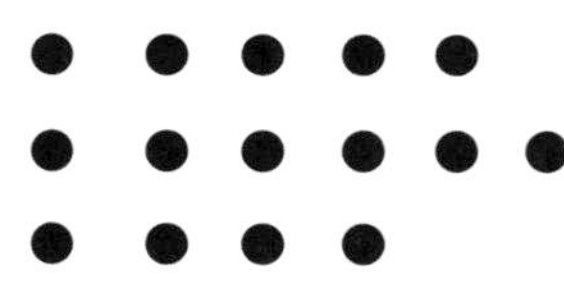
Fig. 3

3.

Fig. 1

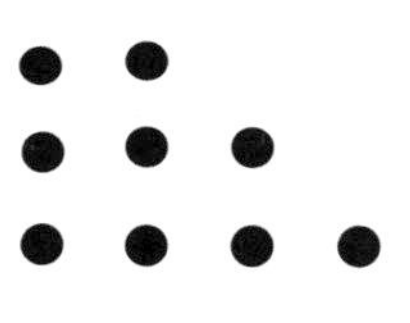
Fig. 2

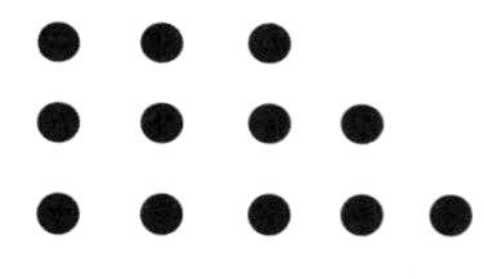
Fig. 3

4.

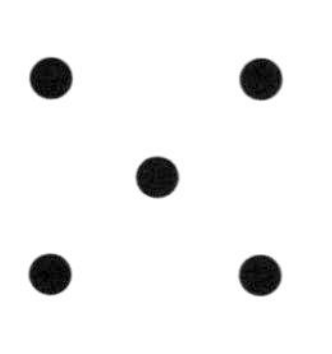
Fig. 1

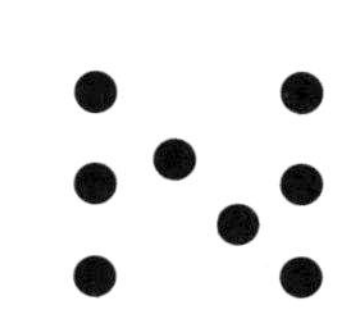
Fig 2

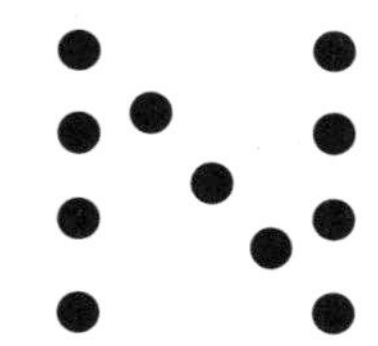
Fig 3

5.

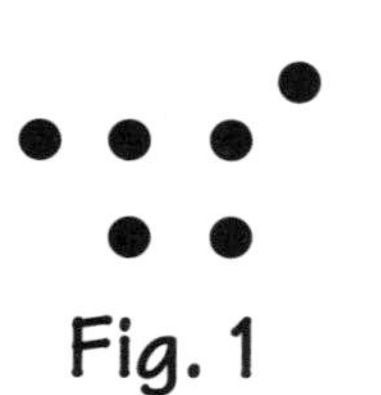
Fig. 1

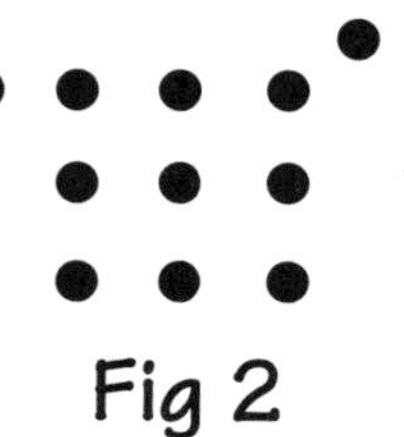
Fig 2

Fig 3

Make up some of your own figures!

HONEYCOMB DECIMALS

Examine this honeycomb shape to find a pattern. Fill in the empty cells with the correct decimals.

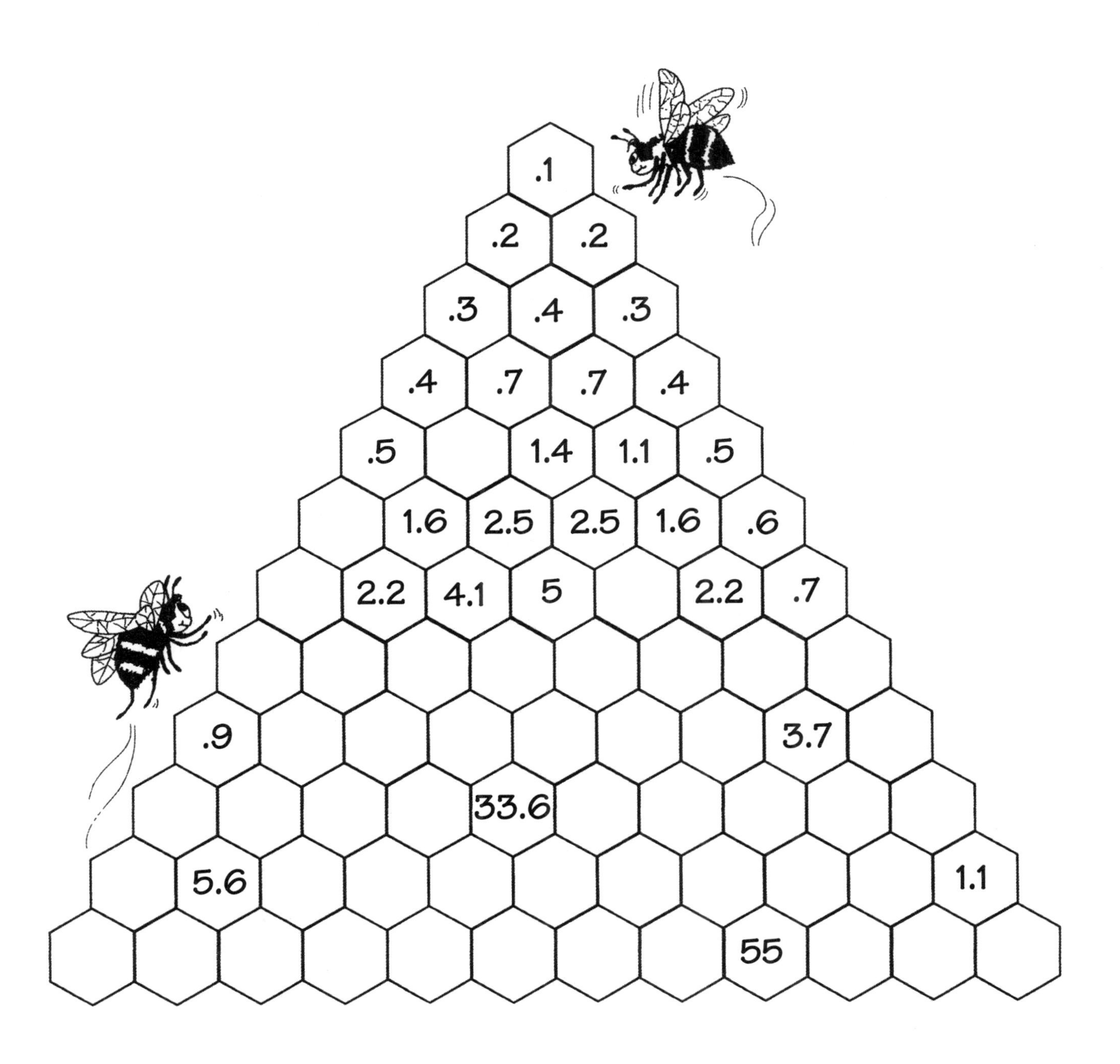

BINARY SORTING CARDS

Make a set of 15 cards as shown on the right, being careful to cut each card by the holes as illustrated. The cards show applications of the binary sequence used in data processing systems.

Discover the pattern in the construction of the cards. On each card, the holes have these values.

8 4 2 1

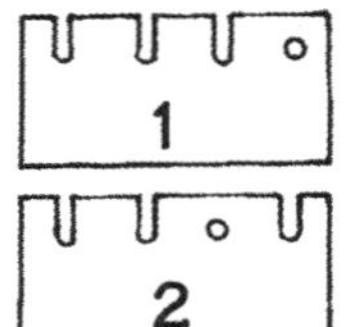

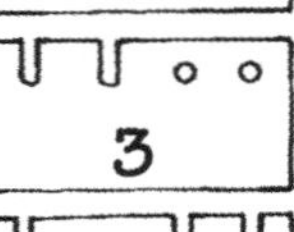

ACTIVITIES WITH BINARY CARDS

1. TASK: Shuffle cards well and place them in a stack. With only four sorts, arrange the cards in numerical order from 1 to 15.

   KEY: Place pen in first hole from the right. Lift cards and place behind stack. Repeat, moving from right to left. When completed, cards will be in numerical order.

2. TASK: Shuffle cards. Pull out all odd-numbered cards with only one sort.

   KEY: Place pen in first hole from the right. Lift cards. These are the odd numbered cards. The cards remaining are the even-numbered cards.

3. TASK: With only four sorts, locate card number 13.

   KEY: Think of the code for card 13. Place pen in first hole from the right and lift cards. 13 is in the lifted set. Place pen in second hole from right in the lifted set. 13 is in the fallen set. Place pen in third hole from the right of the fallen set. 13 is in the lifted set. Place pen in fourth hole from the right of the lifted set. 13 is the only card lifted. Any card in the set can be isolated in four sorts by thinking about the card's code.

4. How many cards would a set contain if each card had
   (a) 1 hole?
   (b) 2 holes?
   (c) 3 holes?
   (d) 4 holes?
   (e) 5 holes?
   (f) 10 holes?
   (g) n holes?

5. Describe the procedure for arranging the cards in descending order from 15 to 1.

ABACUS ABRACADABRA

Study the numbers on these drawings of an abacus. Use what you observe to complete the activity.

1

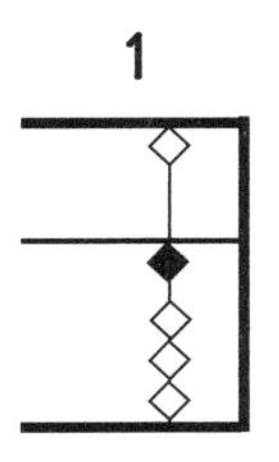

2

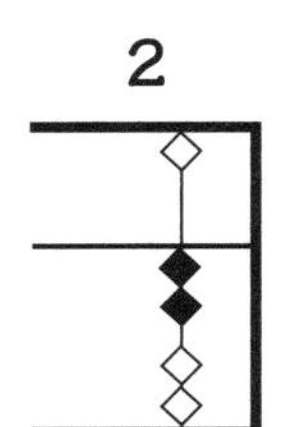

3

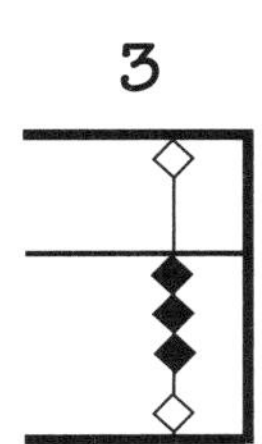

4

5

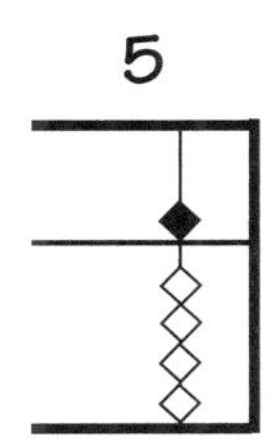

6

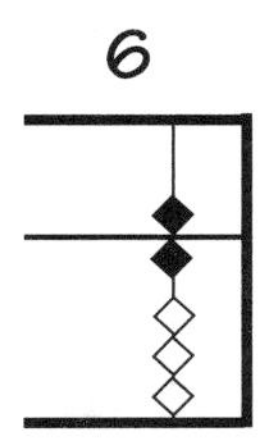

7

8

9

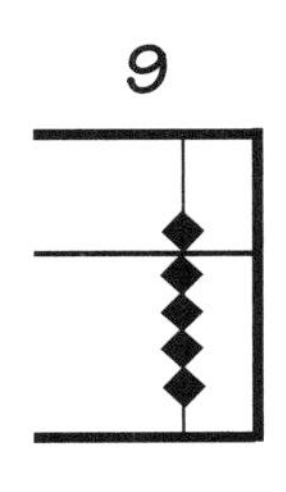

10

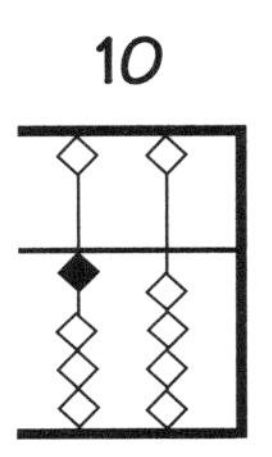

11

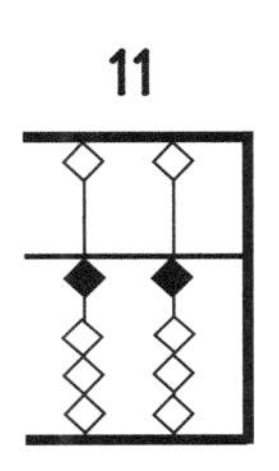

12

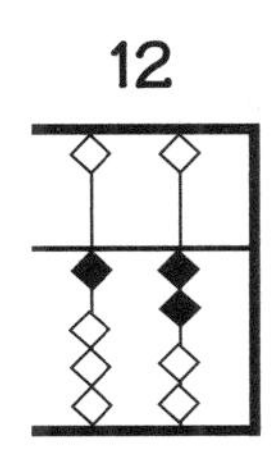

13

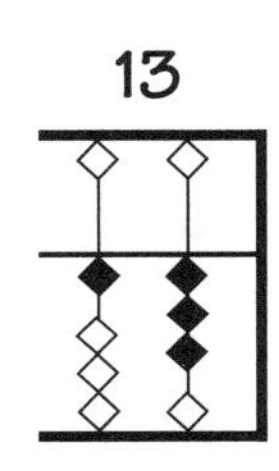

14

Shade the beads you would move to create these numbers.

15

26

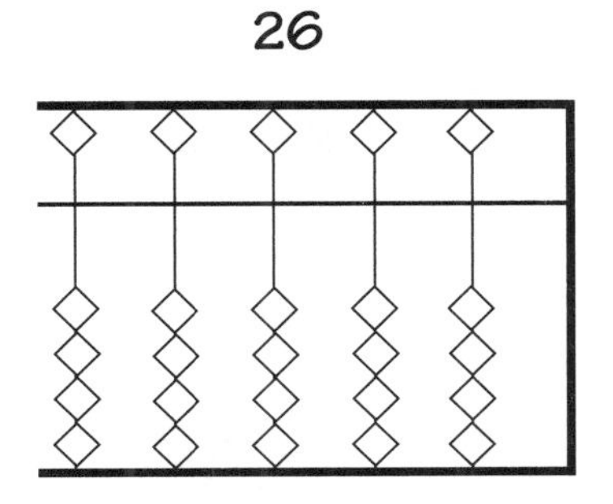

467

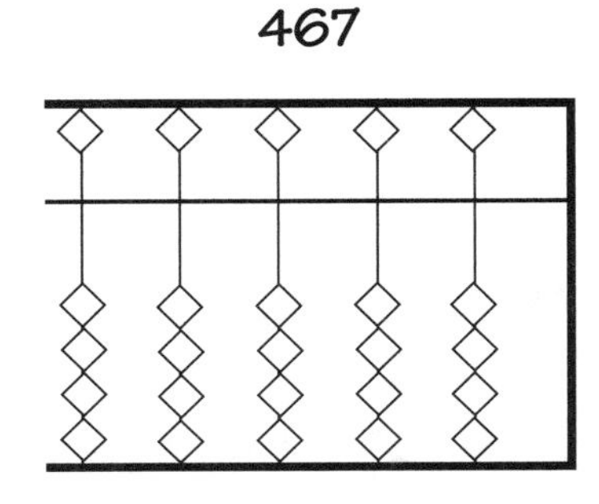

826

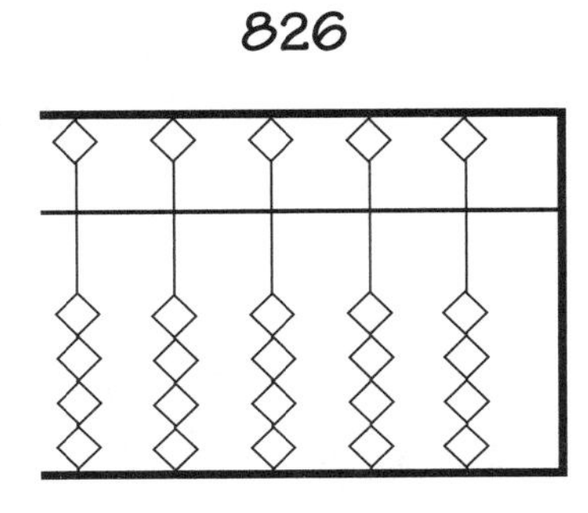

1364

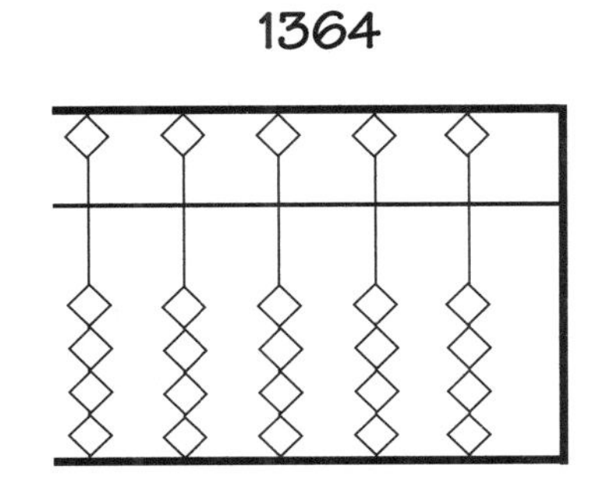

17829

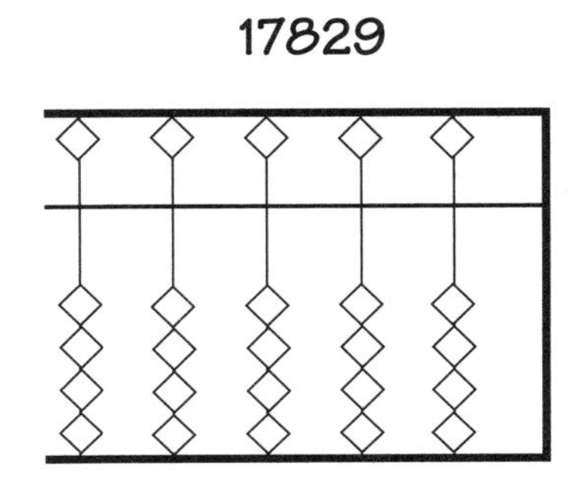

THE MISSING LINK

These numbers in ordered pairs are related to each other. Find what holds these pairs together and fill in the missing links.

1. (4, 6), (12, 18), (2, 3), (6, 9), (20, ___), (8, ___), (10, ___)

2. (6, 4), (4, 3), (10, 6), (12, 7), (2, ___), (14, ___), (___ , 19)

3. (6, 2), (9, 3), (18, 6), (24, ___), (36, ___), (___ , 10)

4. (2, 7), (4, 13), (5, 16), (1, ___), (7, ___), (n, ___)

5. (39, 13), (24, 8), (42, ___), (21, ___), (27, ___)

6. (6, 42), (4, 28), (2, ___), (10, ___), (___ , 56)

7. (5, 24), (3, 8), (7, 48), (2, ___), (10, ___), (20, ___)

8. (12, 9), (20, 17), (6, 3), (15, ___), (30, ___), (___ , 18)

9. (3, 5), (5, 9), (10, 19), (6, ___), (___ , 7), (13, ___)

10. (4, 14), (2, 8), (6, 20), (3, ___), (___ , 38), (10, ___)

11. (18, 2), (9, 1), (36, 4), (63, ___), (___ , 6), (___, n)

12. (4, 27), (2, 13), (5, 34), (7, 48), (3, ___), (6, ___), (___, 62)

EXTRA CHALLENGES:

1. (2, 7), (6, 15), (5, 13), (10, ___), (15, ___), (8, ___)

2. (3, 12), (7, 56), (5, 30), (8, ___), (2, ___), (10, ___)

GIVE ME FIVE !

With only seconds left, the forward rushes in for a lay up and scores! Euclid High School has won its first basketball championship! Celebrating their victory, all nine players run around the court, giving each other "high fives."

If each Euclid player gave each teammate a high five, how many high fives would be given altogether?

Complete the table to solve the problem.

Number of Players	Number of High Fives
2	1
3	
4	
5	
6	
7	
8	
9	
n	

THE UPS AND DOWNS

Study the following equations and write the next five.

$$121 = \frac{22 \times 22}{1+2+1}$$

$$12321 = \frac{333 \times 333}{1+2+3+2+1}$$

$$1234321 = \frac{4444 \times 4444}{1+2+3+4+3+2+1}$$

UP BY THREE

3 + 6 + 9 + 12 + ...

What is the sum of the first 50 terms in this series?

Solve this problem by beginning with an easier problem. Find the sum of one term, two terms, three terms, etc.

Record your results in the table. When you find a pattern, use it to complete the table.

3 = 3
3 + 6 = 9
3 + 6 + 9 = ?
3 + 6 + 9 + 12 = ?

Number of Terms	Sum
1	3
2	9
3	
4	
5	
6	
7	
50	
n	

TINY TRIANGLES

Find the total number of tiny triangles in the 80th triangular number.

To solve this problem, first count the number of triangles created in the 2nd triangular number. Continue through the next triangular numbers until you see the pattern developing.

Record your findings in the table.

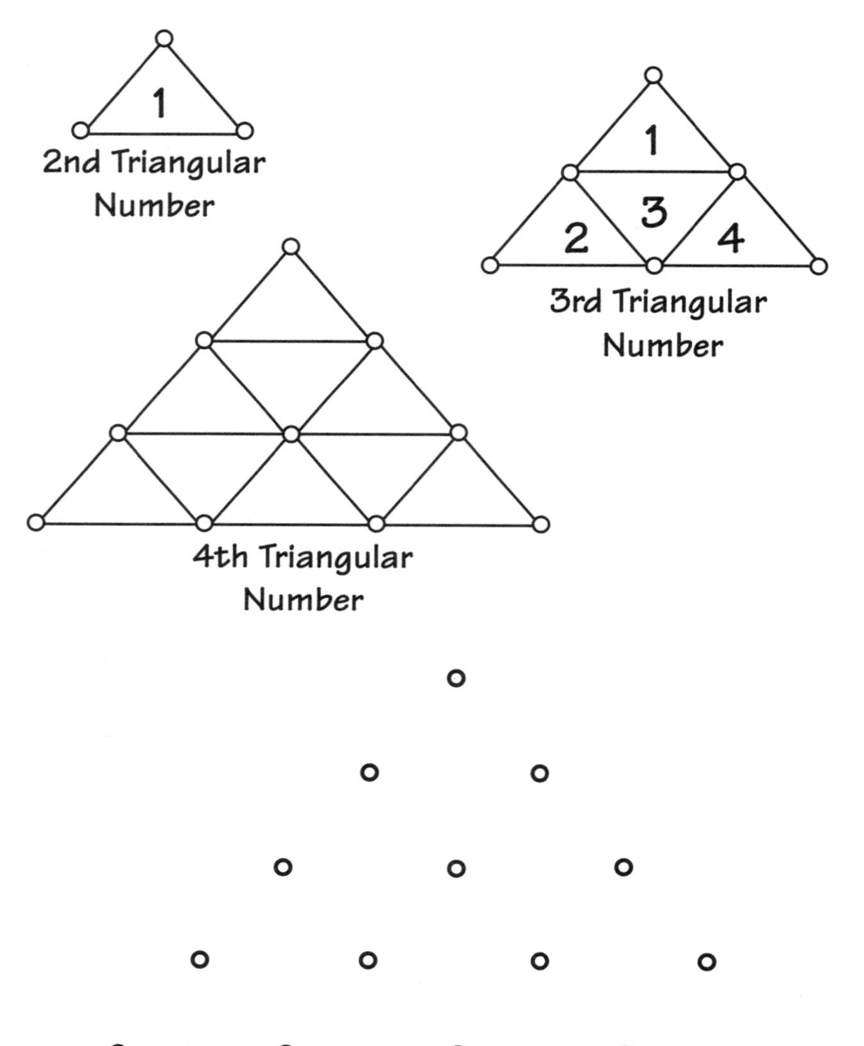

Triangular Number	Number of Triangles
2nd	1
3rd	4
4th	
5th	
6th	
7th	
8th	
9th	
80th	
nth	

BUILDING BLOCKS

Block upon block, layer upon layer, patterns are developing. Complete the tables to discover the "blueprint" for each building. Use your discovery to predict the number of blocks needed for the 50th and nth buildings.

Building Number	Number of Blocks Needed
1	1
2	
3	
4	
5	
6	
50	
n	

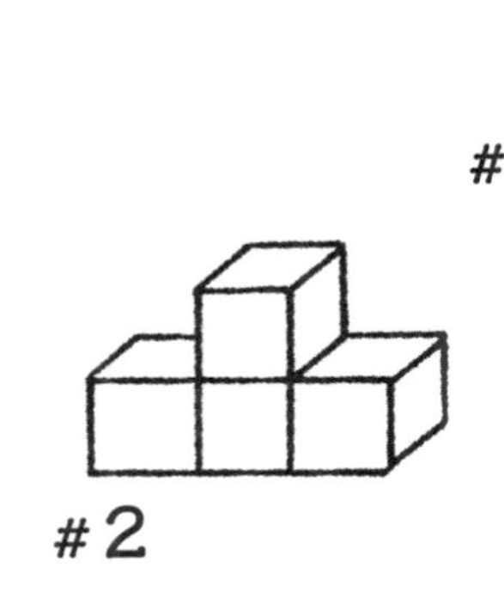

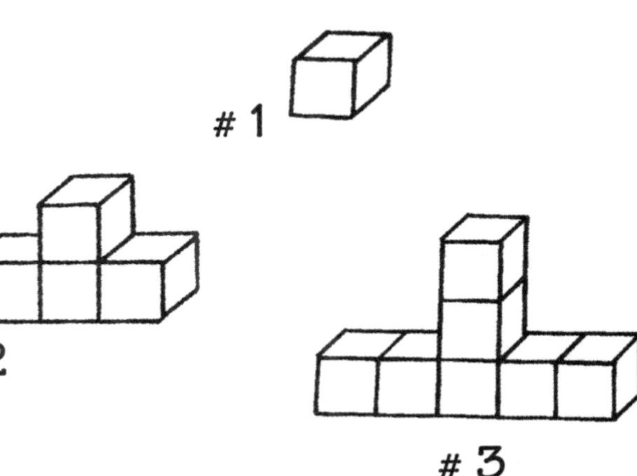

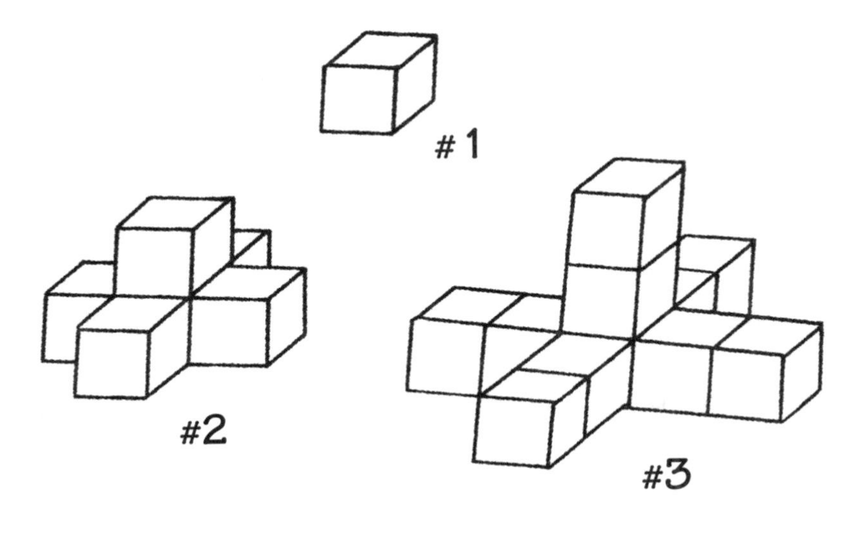

Building Number	Number of Blocks Needed
1	1
2	
3	
4	
5	
6	
50	
n	

FOREVER AND EVER, AMEN !

$$\frac{1}{1} + \frac{1}{2} + \frac{1}{4} + \frac{1}{8} + \frac{1}{16} + \frac{1}{32} + \frac{1}{64} + \cdots$$

What is the sum of the first 20 terms in this series?

Finding the sum of a series often becomes easier by looking at simpler cases. Instead of trying to add all 20 terms, add the first term, then the first two, the first three, and so on. Continue until you find a pattern.

This pattern will help you solve the problem!
Use the table to record your results.

Number of Terms	Sum
1	1
2	$1\frac{1}{2}$
3	
4	
5	
6	
7	
20	

Suppose the above series is continued forever. Do you think the sum will ever be equal to 2? Why or why not?

HINGED !

Use the space below to draw additional hinged shapes and record their perimeters. For example, one triangle has a perimeter of three units. When an identical triangle is "hinged"onto the first, the perimeter becomes four units.

Triangles:

1 1
1

1 1
1
1

Squares:

Pentagons:

Hexagons:

Use the patterns you discover to complete this table.

Number of Polygons	Perimeter of Triangles	Perimeter of Squares	Perimeter of Pentagons	Perimeter of Hexagons
1	3	4	5	6
2	4			
3				
4				
5				
10				
n				

EXTRA CHALLENGE:
What rule could be used to find the perimeter of n hinged polygons when each polygon has s sides?

THE PERFECT SHUFFLE

Split a deck of six cards in half. Shuffle the cards by alternately choosing one from each of the two halves so that the card that was originally on top remains on top after each shuffle.

How many shuffles will return the deck to its original order?

Consider this deck of six cards represented as ABCDEF.

Original Order	After 1 Shuffle	After 2 Shuffles	After 3 Shuffles	After 4 Shuffles
A	A	A	A	A
B	D	E	C	B
C	B	D	E	C
D	E	C	B	D
E	C	B	D	E
F	F	F	F	F

A deck of six cards returns to its original order after only four shuffles!

Using the process outlined above, determine how many shuffles would return decks of four, eight, and ten cards to their original orders. Record your results in the table.

Number of Cards	Number of Shuffles
4	
6	4
8	
10	

THE SHUFFLE FORMULA

How many perfect shuffles are required to bring a deck of 52 cards to its original order? What is your guess?

There is an easy way to find the answer!

Number of Shuffles	Find Smallest Number C-1 Divides Into
1	1
2	3
3	7
4	15
5	31
6	63
7	127
8	255
9	511
10	1023
11	2047
12	4095

Let C represent the number of cards in a deck. Find the smallest number in the second column of the table that C-1 divides into evenly. The corresponding number in the 1st column tells how many shuffles are needed.

A deck of six cards, for example, requires a minimum of four shuffles. In this case C-1 = 6-1 = 5. Five divides into 15, which corresponds to four shuffles.

Use the above process to complete this table.

Number of Cards in Deck	Number of Shuffles to Bring Deck to Original Order
4	
6	4
8	
10	
12	
14	
16	
22	
32	
52	

Was your guess for 52 cards close?

What patterns can you find?

SIDESWIPED SQUARES

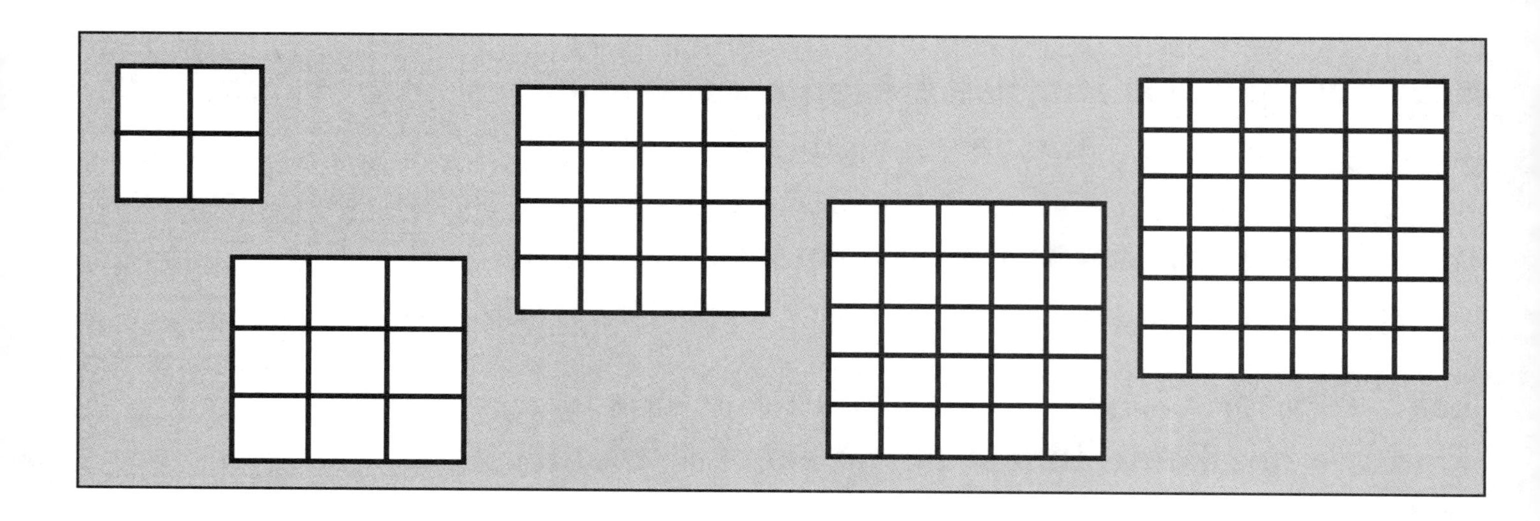

Each of the larger squares is made up of unit squares. Some of these unit squares touch the shaded area on two sides, some on one side only, and some on no sides.

Complete the table using the squares pictured.

Length of One Side	Area of Square	Number of Unit Squares Touching On		
		2 Sides	1 Side	No Sides
2	4	4	0	0
3	9	4	4	
4				
5				
6				
7				
10				
n				

HARE CITY POPULATION EXPLODES !

On January 1 the current population of Hare City is one newly born pair of rabbits (male and female). When they are exactly two months old, this pair will produce a new pair of rabbits (male and female). Assume this pair continues producing a pair of rabbits each month thereafter.

If each pair of rabbits starts producing when they are exactly two months old, and then continues producing a pair each month thereafter, determine the population of Hare City after 12 months.

To help you solve this problem, complete the table, taking advantage of the pattern that emerges.

Jan (A)

Feb (A)

Mar (A)—(B)

Apr (C)—(A) (B)

May (C) (A)—(D) (E)—(B)

Jun (G)—(C) (F)—(A) (D) (E) (B)—(H)

= pair, mature enough to produce

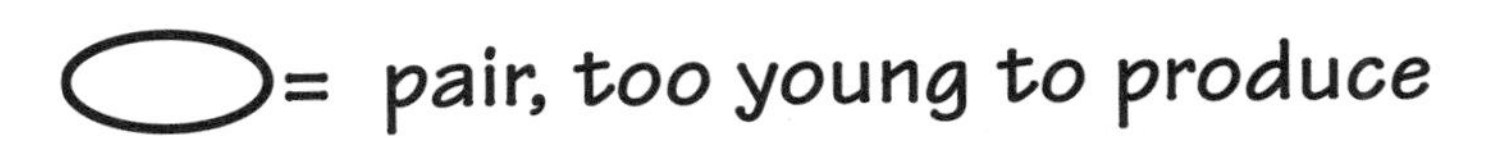

= pair, too young to produce

Month	Number of Pairs
Jan	1
Feb	1
Mar	2
Apr	
May	
Jun	
Jul	
Aug	
Sep	
Oct	
Nov	
Dec	

CHAIN LETTER MADNESS

One day Jennifer received this letter in the mail:

Dear Jennifer,

This is your lucky day! If you follow the instructions in this letter you will receive thousands of dollars in the mail!

Here's what you do:

First, send $1.00 to the person at the top of the list of names below. Take his/her name off and move all the other names up one spot, adding yours in the number 4 position. Then make 20 copies of this new letter and send them to 20 of your best friends.

1. Suzanne Wong
2. Diego Gomez
3. Jason Smith
4. Allison Hunter

P.S. Don't be a bad sport and break the chain, and don't tell the Postal Service about this letter (it's illegal)!

How much would Jennifer get if the chain were not broken? $160,000.
This opportunity sounded too good to be true, and it was! Why?

Complete the table to find how many people would be needed to keep this letter going through only eight cycles. How does your result compare with the current world's population of about five billion?

Cycle Number	Number of People Needed for Cycle	Total Number of People Needed from Beginning
1	20	20
2	400	420
3	8000	8420
4		
5		
6		
7		
8		

PATTERNS WITH A POINT

Every rational number (a fraction made of two integers) becomes a decimal with a pattern when the denominator is divided into the numerator.

Use the pattern to predict the 50th digit for each decimal representation.

a. $\frac{5}{11} = .545454...$

Decimal Place	Digit
1st	5
2nd	4
3rd	5
4th	4
5th	5
6th	4
50th	

b. $\frac{7}{33} = .212121...$

Decimal Place	Digit
1st	2
2nd	1
3rd	2
4th	1
5th	2
6th	1
50th	

c. $\frac{5}{27} = .185185...$

Decimal Place	Digit
1st	1
2nd	8
3rd	5
4th	1
5th	8
6th	5
50th	

Make your own tables to predict the 50th decimal digit of each of these rational numbers.

d. $\frac{23}{99}$ e. $\frac{17}{99}$ f. $\frac{8}{27}$ g. $\frac{3}{11}$

MAGIC MULTIPLICATION

Here are some multiplication problems in which the tens digits are the same and the units digits add up to ten.

See if you can discover a short-cut for finding these products.

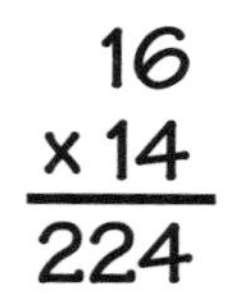

16	27	48	19	62
x 14	x 23	x 42	x 11	x 68
224	621	2016	209	4216

What do you observe about the last two digits in each product? Now examine all but the last two digits in each product. What do you see?

Use what you have discovered to predict the products for these problems.

63	56	78	51	24
x 67	x 54	x 72	x 59	x 26

92	85	43	71	34
x 98	x 85	x 47	x 79	x 36

PIERCED POLYGONS

A *pierced polygon* is a polygon with a polygon-shaped hole.

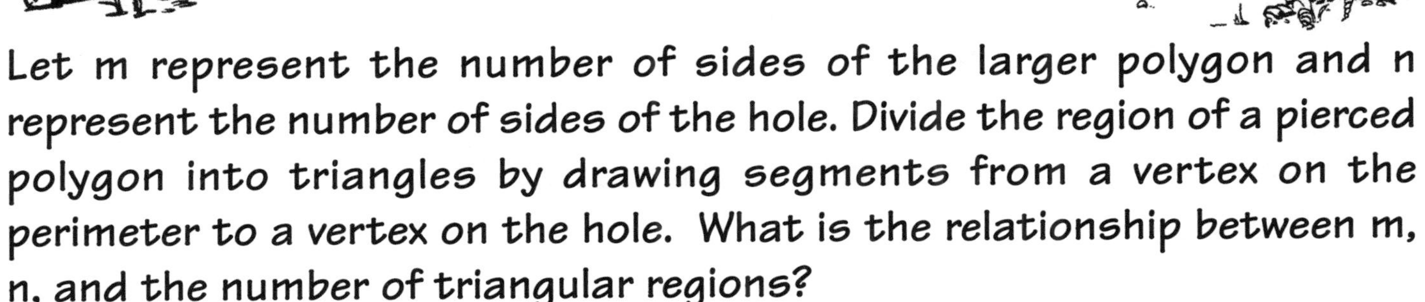

Let m represent the number of sides of the larger polygon and n represent the number of sides of the hole. Divide the region of a pierced polygon into triangles by drawing segments from a vertex on the perimeter to a vertex on the hole. What is the relationship between m, n, and the number of triangular regions?

Discover the relationship by completing the table.

Figure	m	n	Number of Triangles
1	4	3	7
2			
3			
4			
5			
6			

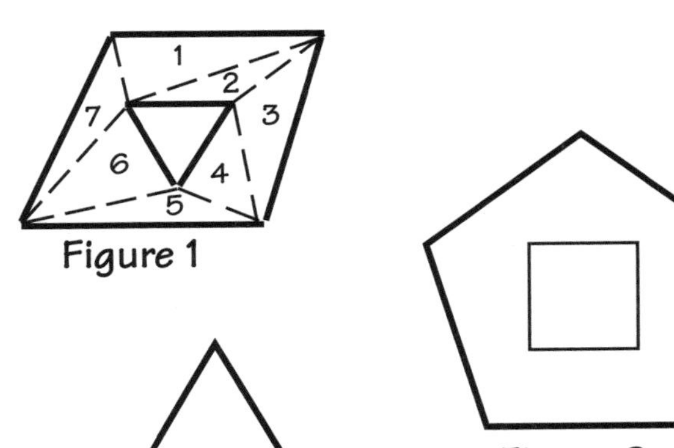

Figure 1

Figure 2

Figure 3

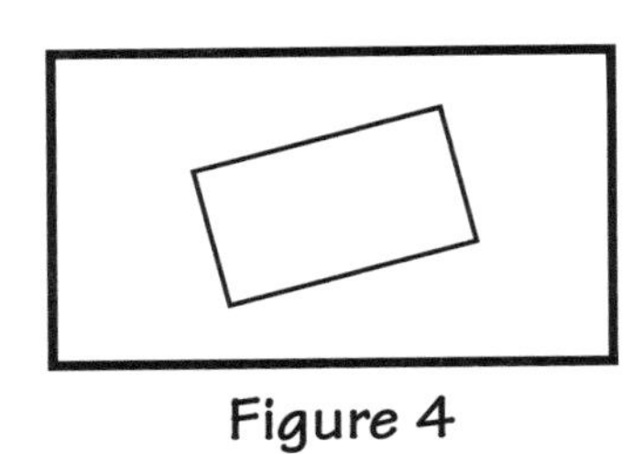

Figure 4

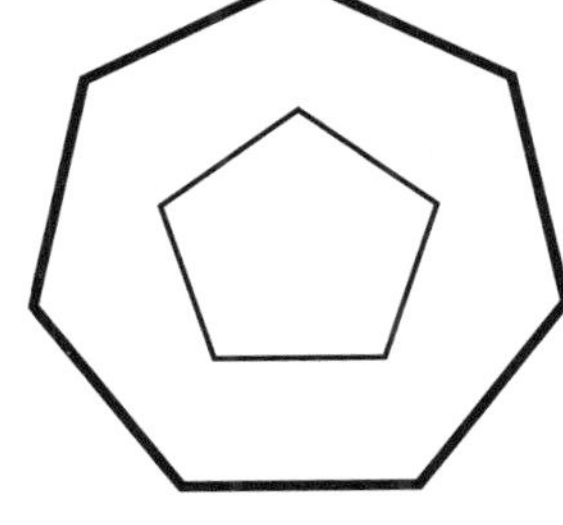

Figure 5

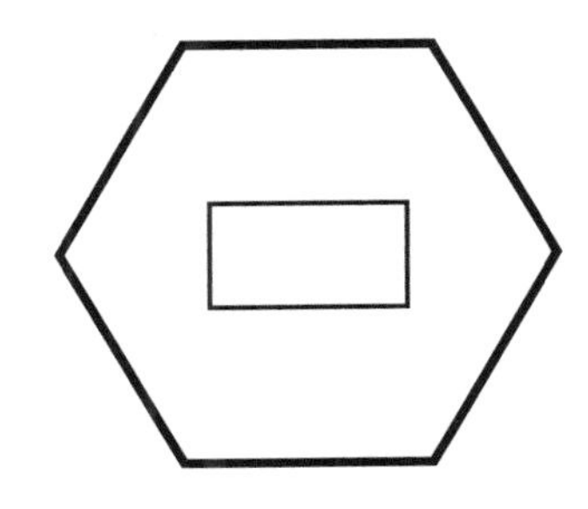

Figure 6

SUM WILL, SUM WON'T

Board games often require players to toss dice. Most players think they're just lucky to roll a certain sum. However, some sums are more likely than others!

Which sum has the greatest chance of occurring when you toss two dice? three dice? 40 dice?

Complete the table to discover an interesting solution.

Number of Dice	Smallest Possible Sum	Largest Possible Sum	Sum(s) With Greatest Chance of Occurring
2	2	12	7
3	3	18	10 or 11
4	4	24	14
5	5	30	17 or 18
6	6	36	21
7	7	42	24 or 25
8	8	48	28
9			
20			
40			
100			
n (even)			
n (odd)			

THE END OF THE WORLD

This pyramid puzzle is called "The Tower of Hanoi" or "End of the World." According to legend, priests in an ancient Hindu temple were given a stack of 64 gold disks, each one a little smaller than the one beneath it. Their assignment was to transfer the 64 disks from one of three poles to another, moving only one disk at a time. A larger disk could never be placed on top of a smaller one. When the priests finished their work, the myth said, the world would vanish.

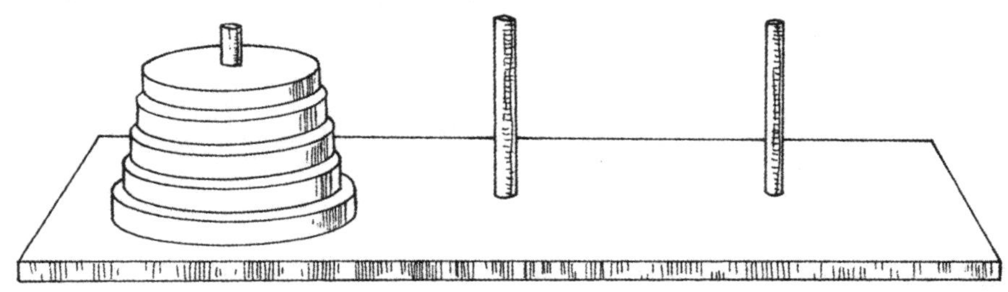

How many moves would be required to transfer the 64 disks one at a time to another pole?

To solve this problem, begin with an easier one. Simulate the moves by using coins of varying sizes and the grid below.

Number of Disks	Minimum Number of Moves
1	1
2	
3	
4	
5	
64	
n	

EXTRA CHALLENGE:

If one disk were moved every second, how many years would it take for all 64 disks to be moved from one pole to another?

CIRCLE CIRCUS

Find the maximum number of interior regions created by 10 intersecting circles.

First determine the maximum number of interior regions possible when fewer circles intersect. Look for the pattern that develops and use it to complete the table.

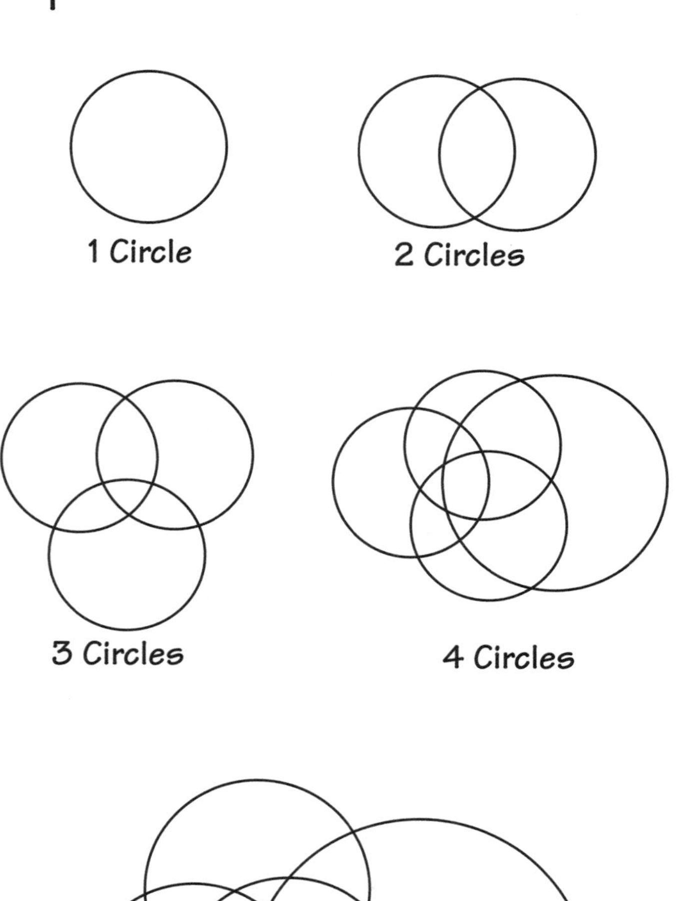

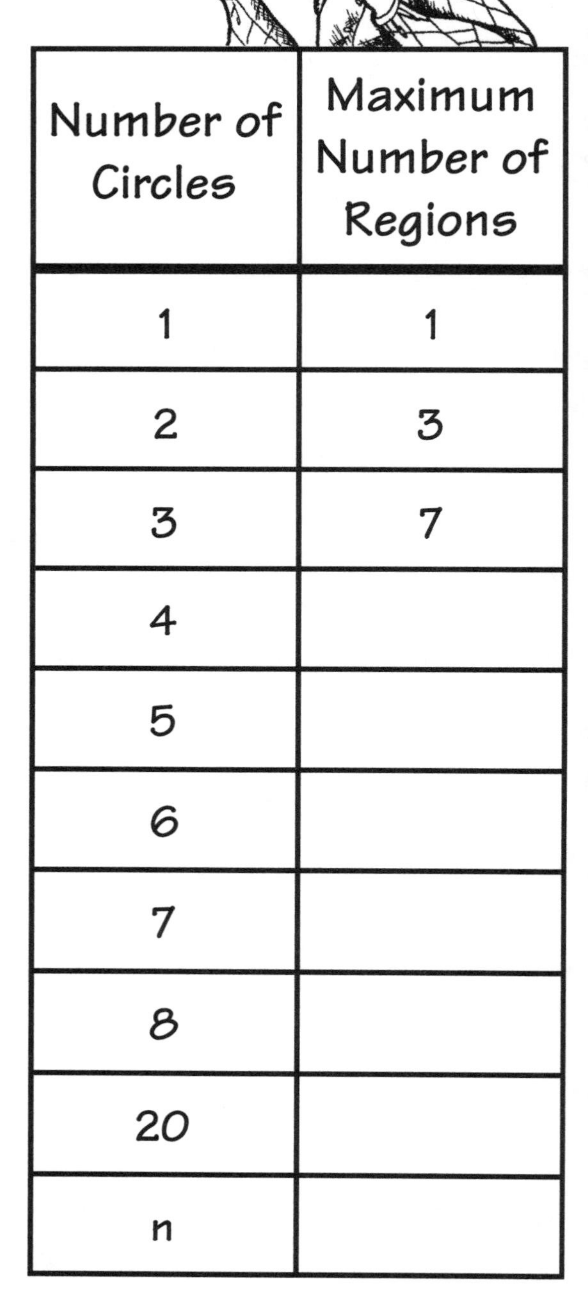

Number of Circles	Maximum Number of Regions
1	1
2	3
3	7
4	
5	
6	
7	
8	
20	
n	

TWENTY-ONE CONNECT

Here is a circle with 21 points joined by all possible line segments. Find the total number of line segments in the drawing by first solving several easier problems.

Count the number of line segments in circles with two points, three points, and so on, and record your findings in the table. Discover the pattern and solve the problem.

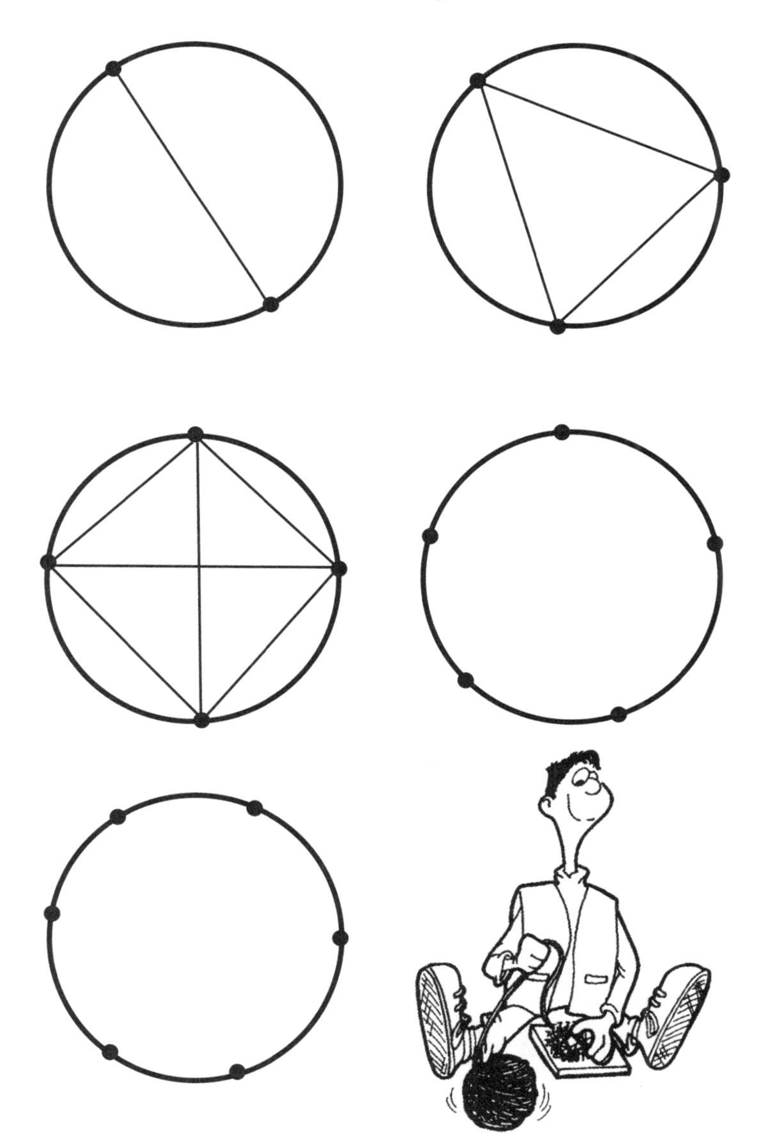

Number of Points	Number of Segments
2	1
3	3
4	
5	
6	
7	
21	
n	

HEXAGONAL NUMBERS

Hexagonal numbers are numbers that can be represented by dots in a hexagonal array. The first four hexagonal numbers are pictured here.

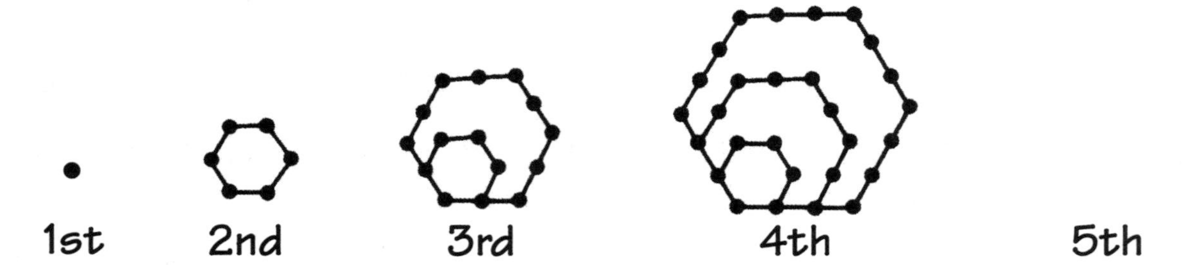

Draw the 5th hexagonal number.

Record the number of dots required to draw each of the first five hexagonal numbers.

Discover a pattern and, without drawing the figures, predict the number of dots required for the 6th, 7th, and 8th hexagonal numbers.

Hexagonal Number	Number of Dots
1st	1
2nd	6
3rd	
4th	
5th	
6th	
7th	
8th	

EXTRA CHALLENGE:
How many dots are needed for the nth hexagonal number?

FRIGHTENING FRACTIONS

$$\frac{1}{1 \cdot 2} + \frac{1}{2 \cdot 3} + \frac{1}{3 \cdot 4} + \frac{1}{4 \cdot 5} + \frac{1}{5 \cdot 6} + \cdots$$

What is the sum of the first 50 terms in this series?

Finding the sum of a series often becomes easier by looking at simpler cases. Instead of trying to add all 50 terms, add the first term, then the first two, the first three, and so on. Continue until you find a pattern.

This pattern will help you solve the problem!
Use the table to record your results.

Number of Terms	Sum
1	$\frac{1}{2}$
2	
3	
4	
5	
6	
7	
50	

EXTRA CHALLENGE:
What is the sum of the first n terms?

MORE TO SEE THAN ONE, TWO, THREE !

Much more hides in this array than the first five counting numbers. Study the array and see what you find.

1

2 2

3 3 3

4 4 4 4

5 5 5 5 5

Discover a pattern in the row sums which would allow you to continue the second column of the table indefinitely *without extending* the array.

Complete the table.

Row Number	Sum
1	1
2	4
3	
4	
5	
6	
7	
n	

EXTRA CHALLENGE:
Find a pattern in the table representing the sum of all numbers in the array.

Number of Rows in Array	Sum of all Numbers in Array	Helping Column
1	1	1 • 2 • 3
2	5	2 • 3 • 5
3	14	3 • 4 • 7
4	30	4 • 5 • 9
5		
6		
7		
n		

ONE POTATO, TWO...

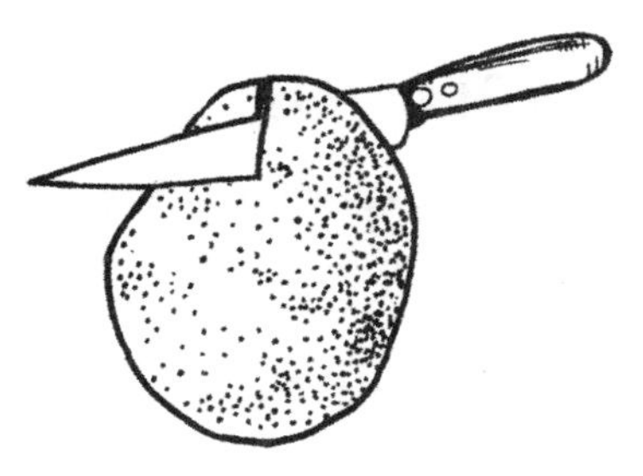

If you make eight straight cuts through one potato, what is the maximum number of pieces possible?

To answer this question, begin by examining an easier problem. Record the number of pieces after zero cuts, one cut, two cuts, etc. Check your results for a pattern. Use this pattern to help you solve for eight cuts. Note that the number of pieces for four cuts is 15, not 16!

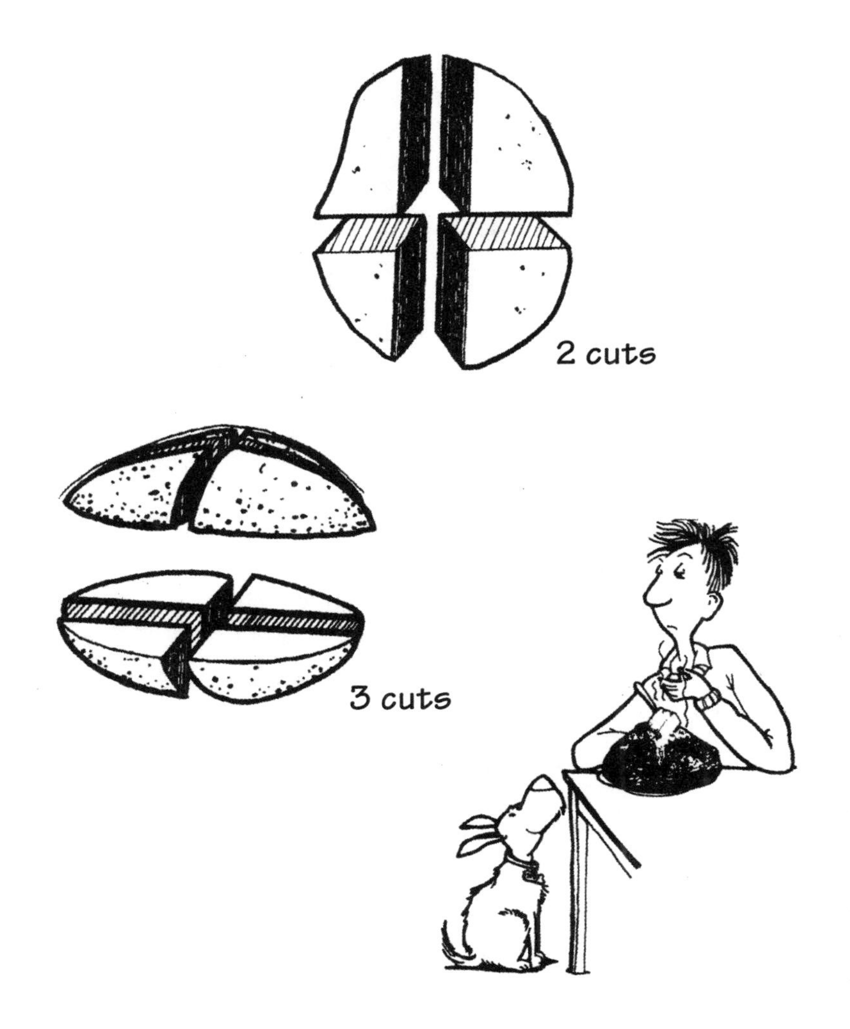

Number of Cuts	Maximum Number of Pieces
0	1
1	2
2	4
3	8
4	15
5	
6	
7	
8	

EXTRA CHALLENGE:
Can you discover how many pieces would result from n cuts?

PASCAL WINS THE WORLD SERIES

In 1993, the Toronto Blue Jays won the World Series. They won the first game, lost the second, won the next two, lost the fifth, and won the sixth. This can be represented by WLWWLW (W=win, L=loss).

The World Series is won by winning the best out of seven games. How many different patterns of wins and losses could occur?

To solve this problem, first look at some easier problems.

Best out of	Number of Possibilities
2	1
3	3
4	4
5	10
6	
7	

Best out of two:
W W
(1 possibility)

Best out of three:
W W
L W W
W L W
(3 possibilities)

Best out of four:
W W W
L W W W
W L W W
W W L W
(4 possibilities)

Best out of five:
W W W
L W W W
W L W W
W W L W
L L W W W
W L L W W
W W L L W
L W L W W
L W W L W
W L W L W
(10 possibilities)

Use Pascal's triangle to find the missing values in the table.

SECRETS WORTH SHARING

Carefully examine the following problems. Each one is unique, yet each hides a pattern that makes the solution simple!

Discover the pattern and you'll have a secret worth sharing!

1. 11 • 11 = 121
   111 • 111 = 12321
   1111 • 1111 = ___ ___ ___ ___ ___ ___ ___
   11111 • 11111 = ___ ___ ___ ___ ___ ___ ___ ___ ___
   111111 • 111111 = ___ ___ ___ ___ ___ ___ ___ ___ ___ ___ ___

2. 1 • 8 + 1 = 9
   12 • 8 + 2 = 98
   123 • 8 + 3 = ___ ___ ___
   1234 • 8 + 4 = ___ ___ ___ ___
   12345 • 8 + 5 = ___ ___ ___ ___ ___
   123456 • 8 + 6 = ___ ___ ___ ___ ___ ___
   1234567 • 8 + 7 = ___ ___ ___ ___ ___ ___ ___
   12345678 • 8 + 8 = ___ ___ ___ ___ ___ ___ ___ ___
   123456789 • 8 + 9 = ___ ___ ___ ___ ___ ___ ___ ___ ___

3. 1 • 9 + 2 = 11
   12 • 9 + 3 = 111
   123 • 9 + 4 = ___ ___ ___ ___
   1234 • 9 + 5 = ___ ___ ___ ___ ___
   12345 • 9 + 6 = ___ ___ ___ ___ ___ ___
   123456 • 9 + 7 = ___ ___ ___ ___ ___ ___ ___

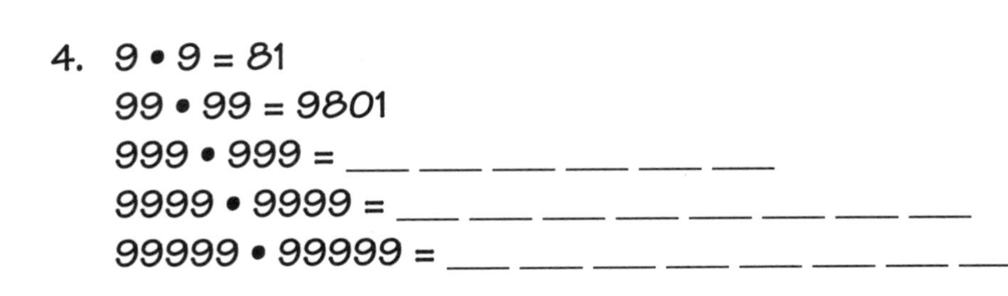

4. 9 • 9 = 81
   99 • 99 = 9801
   999 • 999 = ___ ___ ___ ___ ___ ___
   9999 • 9999 = ___ ___ ___ ___ ___ ___ ___ ___
   99999 • 99999 = ___ ___ ___ ___ ___ ___ ___ ___ ___ ___

THIS LINE'S BUSY !

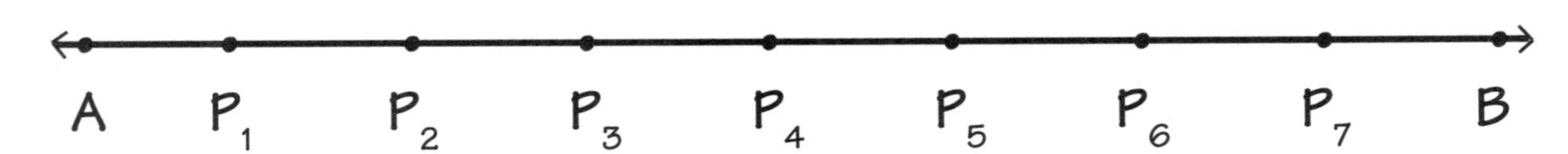

What is the total number of segments on this line?

To help you answer this question, note segment AB with no interior points on the line.

A B

There is one line segment, AB.

Insert one interior point, P_1, on segment AB.

A P_1 B

There are now three segments, AP_1, AB, P_1B.

Complete the table using the diagrams. Find a pattern and predict the number of segments for five, six, and seven interior points.

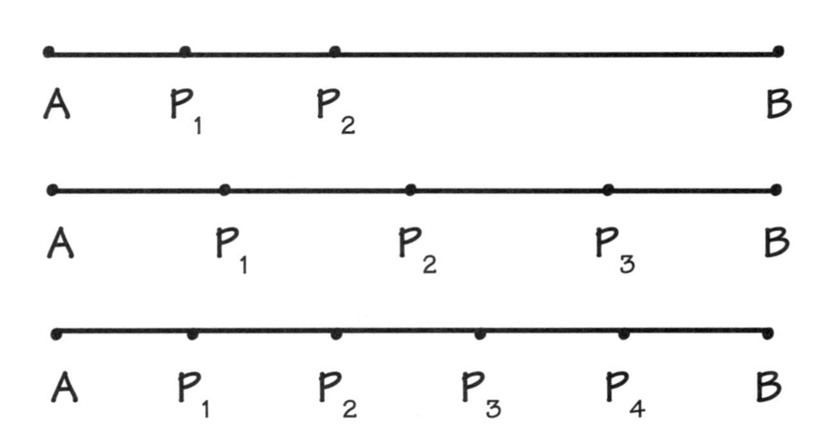

Number of Interior Points	Total Number of Segments
0	1
1	3
2	
3	
4	
5	
6	
7	

EXTRA CHALLENGE:
How many segments are created with n interior points?

THE TWELVE DAYS OF CHRISTMAS

On the twelfth day of Christmas my true love gave to me twelve drummers drumming, eleven pipers piping, ten lords a-leaping, nine ladies dancing, eight maids a-milking, seven swans a-swimming, six geese a-laying, five gold rings, four calling birds, three French hens, two turtle doves, and a partridge in a pear tree.

What is the total number of gifts given on the 12th day? Complete the table to discover how many gifts were given on each of the twelve days of Christmas.

DAY	TOTAL NUMBER OF GIFTS GIVEN
1st	1
2nd	3
3rd	
4th	
5th	
6th	
7th	
8th	
9th	
10th	
11th	
12th	

EXTRA CHALLENGE:
Can you discover the rule to determine how many gifts were given on the nth day?

SAVE THE BEST FOR LAST !

What is the final digit of 9^{12314}?

By looking at some simpler cases and discovering a pattern, this problem becomes easy!

$9^1 = 9$

$9^2 = 81$

$9^3 = 729$

$9^4 =$ ______

$9^5 =$ ______

$9^6 =$ ______

Exponent of 9	Last Digit
1	9
2	1
3	9
4	
5	
6	
12314	

Follow this process and create similar tables to determine the last digit of each of these numbers.

1. 4^{1289}

2. 6^{5633}

3. 5^{823}

APPENDIX

PASCAL'S TRIANGLE

THE SOURCE OF MANY PATTERNS

1

1 1

1 2 1

1 3 3 1

1 4 6 4 1

1 5 10 10 5 1

1 6 15 20 15 6 1

1 7 21 35 35 21 7 1

1 8 28 56 70 56 28 8 1

1 9 36 84 126 126 84 36 9 1

1 10 45 120 210 252 210 120 45 10 1

1 11 55 165 330 462 462 330 165 55 11 1

SQUARE, OBLONG, AND TRIANGULAR NUMBERS

Square numbers are numbers that can be represented by dots in a square array. The first four square numbers are pictured below.

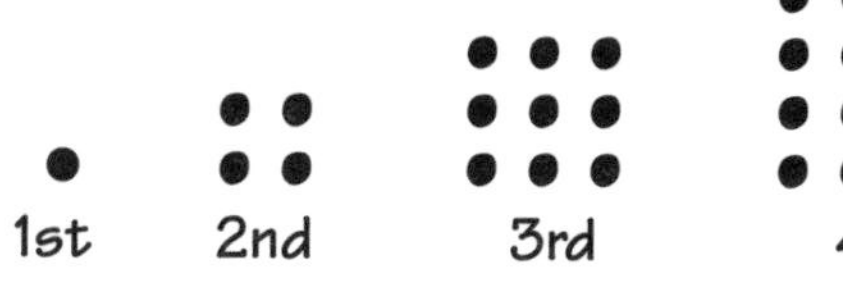

Square Number	Number of Dots
1st	1
2nd	4
3rd	9
4th	16
5th	25
6th	36
50th	2500
nth	n^2

Oblong numbers are numbers that can be represented by dots in a rectangle having one dimension one unit longer than the other. The first four oblong numbers are pictured below.

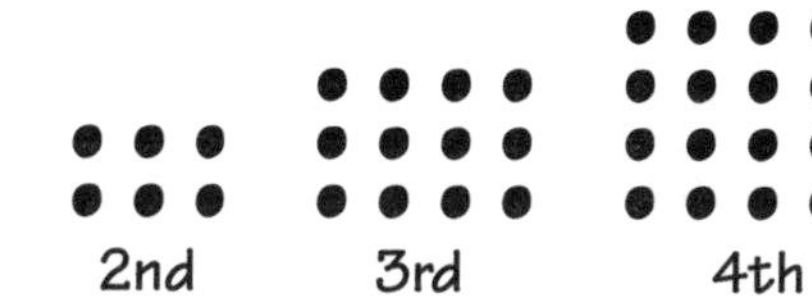

Oblong Number	Number of Dots
1st	2
2nd	6
3rd	12
4th	20
5th	30
6th	42
50th	2550
nth	$n(n+1)$

Triangular numbers are numbers that can be represented by dots in a triangular array. The first four triangular numbers are pictured below.

1st 2nd 3rd 4th

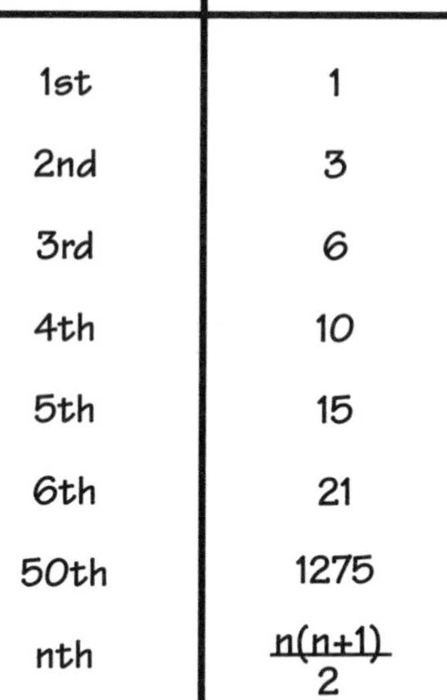

Triangular Number	Number of Dots
1st	1
2nd	3
3rd	6
4th	10
5th	15
6th	21
50th	1275
nth	$\frac{n(n+1)}{2}$

THE METHOD OF FINITE DIFFERENCES

For some of the extra challenges in *What's Next?* the horizontal solutions are fairly complex. An effective approach to these problems is the method of finite differences. This method may be applied to find a horizontal solution (general formula) whenever a *constant* difference occurs between numbers in a sequence.

After data are collected in a table, determine how many columns of differences are required to obtain a column of constant differences.

n			1st diff.		2nd diff.
0	1				
		>	3		
1	4			>	4
		>	7		
2	11			>	4
		>	11		
3	22			>	4
		>	15		
4	37			>	4
		>	19		
5	56				
n	?				

In the example, note that it takes two columns of differences to obtain constants.

The numbers in the first column of differences are 3, 7, 11, 15, and 19. Since these are not constant, create a second column of differences. The differences between these numbers *are* constant, since each difference is 4.

It may be necessary to create several columns of differences to find a constant difference. Once the constant difference has been found, use the formulas in the Finite Difference Chart. Note that there are four tables of formulas on this chart. The chart shows that every first degree expression has only one column of differences; every second degree expression has two columns of differences, etc. The choice of table is determined by how many columns of differences are required to generate a constant difference. For example, if only one column is required, the solution may be found through the first degree table. If three columns are required, the third degree table would be used.

In the earlier example, two columns of differences are required to reach a constant difference so the solution is available through the second degree table. To solve this problem, assign the values in the problem's table to the corresponding expressions in the Finite Difference Chart for second degree expressions.

		1st diff.		2nd diff.	n	an^2+bn+c				
(1)					0	(c)				
	>	(3)						(a+b)		
4			>	(4)	1	a+b+c			>	(2a)
	>	7					>	3a+b		
11			>	4	2	4a+2b+c			>	2a
	>	11					>	5a+b		
22			>	4	3	9a+3b+c			>	2a
	>	15					>	7a+b		
37			>	4	4	16a+4b+c			>	2a
	>	19					>	9a+b		
56					5	25a+5b+c				

Assigning these values creates the following three equations:

$$c = 1$$
$$a + b = 3$$
$$2a = 4$$

Solving these three equations, beginning with $2a = 4$, produces the following values for a, b, and c.

$$a = 2$$
$$b = 1$$
$$c = 1$$

When these values are substituted in the general second degree expression, an^2+bn+c, the result is $2n^2+n+1$, the general formula for the horizontal solution.

While the method of finite differences is frequently a very convenient problem-solving approach, it will not always lead to a solution (constant differences may not occur), nor is it necessarily the best way to approach a problem.

The Finite Difference Chart may be especially useful if students are challenged to discover how the chart is constructed and to experiment with the technique on problems which may have been solved through other processes.

FINITE DIFFERENCE CHART

First Degree

n	an+b	
0	b	
		a
1	a+b	
		a
2	2a+b	
		a
3	3a+b	
		a
4	4a+b	
		a
5	5a+b	

Second Degree

n	an^2+bn+c		
0	c		
		a+b	
1	a+b+c		2a
		3a+b	
2	4a+2b+c		2a
		5a+b	
3	9a+3b+c		2a
		7a+b	
4	16a+4b+c		2a
		9a+b	
5	25a+5b+c		

Third Degree

n	an^3+bn^2+cn+d			
0	d			
		a+b+c		
1	a+b+c+d		6a+2b	
		7a+3b+c		6a
2	8a+4b+2c+d		12a+2b	
		19a+5b+c		6a
3	27a+9b+3c+d		18a+2b	
		37a+7b+c		6a
4	64a+16b+4c+d		24a+2b	
		61a+9b+c		
5	125a+25b+5c+d			

Fourth Degree

n	$an^4+bn^3+cn^2+dn+e$				
0	e				
		a+b+c+d			
1	a+b+c+d+e		14a+6b+2c		
		15a+7b+3c +d		36a+6b	
2	16a+8d+4c+2d+e		50a+12b+2c		24a
		65a+19b+5c+d		60a+6b	
3	81a+27b+9c+3d+e		110a+18b+2c		24a
		175a+37b+7c+d		84a+6b	
4	256a+64b+16c+4d+e		194a+24b+2c		24a
		369a+61b+9c+d		108a+6b	
5	625a+125b+25c+5d+e		302a+30b+2c		
		671a+91b+11c+d			
6	1296a+216b+36c+6d+e				

SOLUTIONS

PRACTICALLY PREDICTABLE!
(pg.1)

1. 30, 25, 20, 15, 10,5
2. January, March, May, July, September, November
3. 6, 4, 2, 0, -2, -4, -6, -8, -10
4. 1/2, 1/4, 1/8, 1/16, 1/32, 1/64
5. z, y, w, t, p, k, e
6. Wednesday, Saturday, Tuesday, Friday, Monday, Thursday
7. 7700, 6600, 5500, 4400, 3300, 2200
8. .125, .25, .375, .5, .625, .75, .875
9. (Uncle, Nephew) , (Aunt, Niece)
10. 1 x 1/2, 2 x 1/4, 3 x 1/6, 4 x 1/8, 5 x 1/10
11. .1, .1, .2, .3, .5, .8, 1.3, 2.1, 3.4

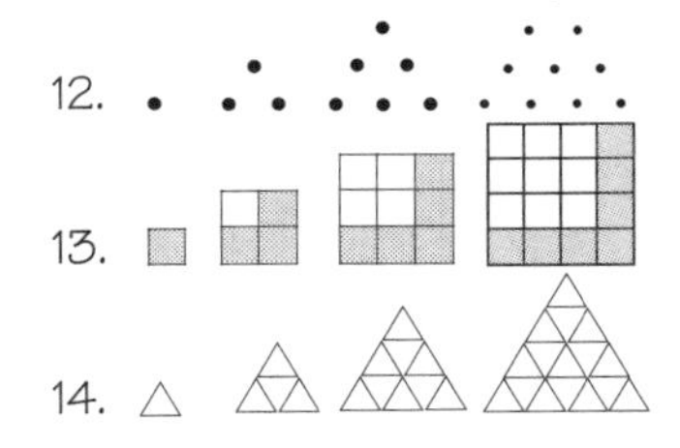

IN AND OUT
(pg.2)

In	Out
math	m
zebra	z
house	h
pick	p
school	s
mouse	m
nose	n

In	Out
guessing	9
involves	9
taking	7
a	2
risk	5
but	4
it	3
is	3
often	6
a	2
good	5
strategy	9

In	Out
problem	13
solving	7
in	14
math	8
can	14
be	5
much	8
fun	14
do	15
you	21
think	11
so	15

In	Out
bath	b
cheetah	f
love	p
line	j
stem	f
elephant	f
sit	j
pin	j
pickle	j
today	p
think	j

In	Out
3	7
10	21
5	11
0	1
50	101
4	9
15	31
6	13
20	41
30	61
100	201

In	Out
2	2
145	10
31	4
10	1
8	8
182	11
0	0
20	2
3	3
481	13
16	7

ROLLER COASTER SUMS
(pg.3)

The sum of each roller coaster series is the square of the highest number in the series. If the highest number is n, the sum in n^2.

SQUARE, OBLONG, AND TRIANGULAR NUMBERS (pg.4)

Square Number	Number of Dots
1st	1
2nd	4
3rd	9
4th	16
5th	25
6th	36
50th	2500
nth	n^2

Oblong Number	Number of Dots
1st	2
2nd	6
3rd	12
4th	20
5th	30
6th	42
50th	2550
nth	n(n+1)

Triangular Number	Number of Dots
1st	1
2nd	3
3rd	6
4th	10
5th	15
6th	21
50th	1275
nth	$\frac{n(n+1)}{2}$

TABLE IT!
(pg.5)

0	8
1	11
2	14
3	17
4	20
5	23
50	158
n	3n+8

1	2
2	7
3	12
4	17
5	22
6	27
50	252
n	5n+2

1	4
2	7
3	12
4	19
5	28
6	39
50	2503
n	n^2+3

0	4
1	9
2	16
3	25
4	36
5	49
50	2704
n	$(n+2)^2$

1	0
2	8
3	16
4	24
5	32
6	40
50	392
n	8n-8 or 8(n-1)

1	1
2	8
3	27
4	64
5	125
6	216
50	125,000
n	n^3

FIGURATIVELY SPEAKING
(pg.6)

Number Family	Hexagonal	Heptagonal	Octagonal
1st	1=1	1=1	1=1
2nd	1+5=6	1+6=7	1+7=8
3rd	1+5+9=15	1+6+11=18	1+7+13=21
4th	1+5+9+13=28	1+6+11+16=34	1+7+13+19=40
5th	1+5+9+13+17=45	1+6+11+16+21=55	1+7+13+19+25=65

CHINESE CHECKERS
(pg.7)

Number of Rows of Holes in Starting Pen	Total Number of Holes
1	13
2	37
3	73
4	121
5	181
6	253
7	337

To discover the vertical pattern in the second column, students should look at the first set of differences—24, 36, 48, 60— which holds the secret to continuing the table indefinitely.

Extra Challenge Solution:
A helpful hint for solving the extra challenge is to subtract 1 from each number in the second column giving 12, 36, 72, 120, etc. From this there are various ways to obtain the general solution $6n(n+1) + 1$.

Additional activity:
Some teachers may wish to provide another interesting activity dealing with the Chinese Checkers board. Ask students to determine the total number of small triangles contained on the boards. The solution to this problem is shown.

Number of Rows of Holes in Starting Pen	Total Number of Small Triangles on Board
1	12
2	48
3	108
4	192
5	300
6	432
n	$12n^2$

Help students find the horizontal solution by suggesting they divide the numbers in the second column by 12. They should discover the perfect squares emerging.

CALENDAR CROSSING
(pg.8)

Entries in students' tables will vary. The sum of the five numbers equals five times the center number regardless of what month or year is selected.

BUSY INTERSECTIONS
(pg.9)

Number of Points	Number of Intersections
4	1
5	5
6	15
7	35
8	70
9	126
10	210

1
1 1
1 2 1
1 3 3 1
1 4 6 4 1
1 5 10 10 5 1
1 6 15 20 15 6 1
1 7 21 35 35 21 7 1
1 8 28 56 70 56 28 8 1
1 9 36 84 126 126 84 36 9 1
1 10 45 120 210 252 210 120 45 10 1
1 11 55 165 330 462 462 330 165 55 11 1

The pattern for this problem appears as a diagonal in Pascal's triangle as shown. A copy of Pascal's triangle appears in the Appendix and should be made available to students.

THE FOLD THAT FOOLS
(pg.10)

Number of Folds	Number of Layers
0	1
1	2
2	4
3	8
4	16
5	32
6	64
50	2^{50}
n	2^n

Extra Challenge Solution: If the piece of paper were folded 50 times, it would be approximately 17,769,885 miles high!

TRIPLE YOUR FUN!
(pg.11)

Triple Pattern #1
(1, 4, 6)
(5, 8, 10)
(0, 3, 5)
(4, 7, 9)
(12, 15, 17)
(5, 8, 10)
(23, 26, 28)
(57, 60, 62)

Triple Pattern #2
(2, 4, 8)
(1, 1, 1)
(4, 16, 64)
(3, 9, 27)
(5, 25, 125)
(0, 0, 0)
(8, 64, 512)
(7, 49, 343)

Triple Pattern #3
(1, 1, 1)
(2, 3, 4)
(3, 6, 9)
(4, 10, 16)
(5, 15, 25)
(6, 21, 36)
(10, 55, 100)
(11, 66, 121)

Triple Pattern #4
(8, 2, 7)
(16, 4, 9)
(28, 7, 12)
(40, 10, 15)
(32, 8, 13)
(12 , 3, 8)
(20, 5, 10)
(44, 11, 16)

Triple Pattern #5
(15, 5, 10)
(27, 9, 18)
(12, 4, 8)
(30, 10, 20)
(9, 3, 6)
(21, 7, 14)
(18, 6 , 12)
(36, 12, 24)

THAT FIGURES!
(pg.12)

1. 400
2. 306
3. 303
4. 304
5. 10303

HONEYCOMB DECIMALS
(pg.13)

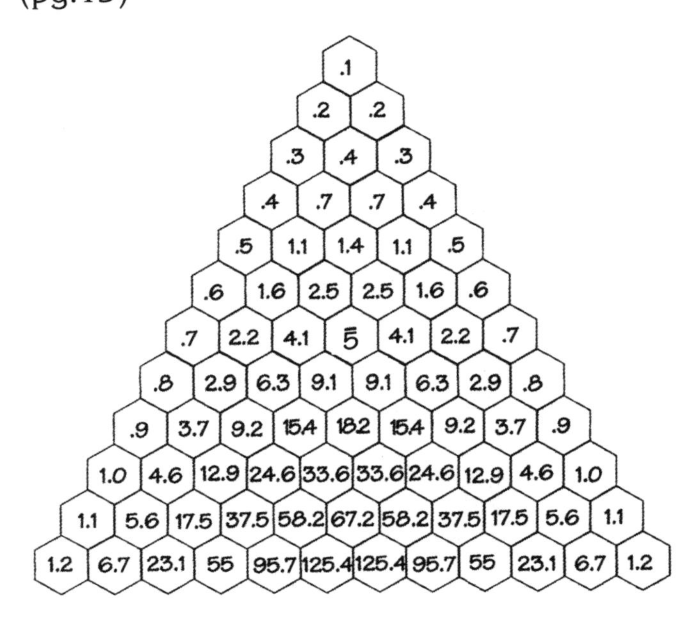

BINARY SORTING CARDS
(pg.14)

Number of Holes	Number of Cards
1	1
2	3
3	7
4	15
5	31
10	1023
n	$2^n - 1$

ABACUS ABRACADABRA
(pg.15)

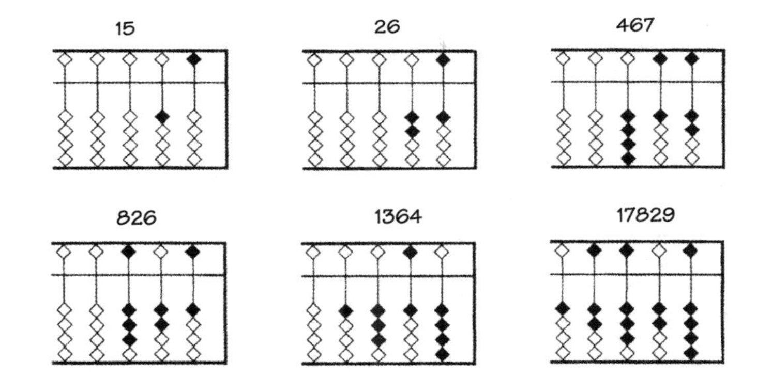

THE MISSING LINK
(pg.16)

1. (4, 6), (12, 18), (2, 3), (6, 9), (20, 30), (8, 12) , (10, 15)
2. (6, 4), (4, 3), (10, 6), (12, 7), (2, 2), (14, 8), (36 , 19)
3. (6, 2), (9, 3), (18, 6), (24, 8), (36, 12), (30, 10)
4. (2, 7), (4, 13), (5, 16), (1, 4), (7, 22), (n, 3n+1)
5. (39, 13), (24, 8), (42, 14), (21, 7), (27, 9)
6. (6, 42), (4, 28), (2, 14), (10, 70), (8 , 56)
7. (5, 24), (3, 8), (7, 48), (2, 3), (10, 99), (20, 399)
8. (12, 9), (20, 17), (6, 3), (15, 12), (30, 27), (21 , 18)
9. (3, 5), (5, 9), (10, 19), (6, 11), (4 , 7), (13, 25)
10. (4, 14), (2, 8), (6, 20), (3, 11), (12 , 38), (10,32)
11. (18, 2), (9, 1), (36, 4), (63, 7), (54 , 6), (9n, n)
12. (4, 27), (2, 13), (5, 34), (7, 48), (3, 20), (6, 41), (9, 62)

EXTRA CHALLENGES:
1. (2, 7), (6, 15), (5, 13), (10, 23), (15, 33), (8, 19)
2. (3, 12), (7, 56), (5, 30), (8, 72), (2, 6), (10, 110)

GIVE ME FIVE!
(pg.17)

Number of Players	Number of High Fives
2	1
3	3
4	6
5	10
6	15
7	21
8	28
9	36
n	$\frac{n(n-1)}{2}$

THE UPS AND DOWNS
(pg.18)

$$121 = \frac{22 \times 22}{1+2+1}$$

$$12321 = \frac{333 \times 333}{1+2+3+2+1}$$

$$1234321 = \frac{4444 \times 4444}{1+2+3+4+3+2+1}$$

$$123454321 = \frac{55555 \times 55555}{1+2+3+4+5+4+3+2+1}$$

$$12345654321 = \frac{666666 \times 666666}{1+2+3+4+5+6+5+4+3+2+1}$$

$$1234567654321 = \frac{7777777 \times 7777777}{1+2+3+4+5+6+7+6+5+4+3+2+1}$$

$$123456787654321 = \frac{88888888 \times 88888888}{1+2+3+4+5+6+7+8+7+6+5+4+3+2+1}$$

$$12345678987654321 = \frac{999999999 \times 999999999}{1+2+3+4+5+6+7+8+9+8+7+6+5+4+3+2+1}$$

UP BY THREE
(pg.19)

The first set of differences for the numbers in the second column is 6, 9, 12, 15, 18, etc. which allows the table to be continued indefinitely.

Number of Terms	Sum
1	3
2	9
3	18
4	30
5	45
6	63
7	84
50	3825
n	$\frac{3n(n+1)}{2}$

Note that dividing each number in the second column by 3 produces the triangular numbers.

TINY TRIANGLES
(pg.20)

Triangular Number	Number of Triangles
2nd	1
3rd	4
4th	9
5th	16
6th	25
7th	36
8th	49
9th	64
80th	6241
nth	$(n-1)^2$

BUILDING BLOCKS
(pg.21)

Building Number	Number of Blocks Needed
1	1
2	4
3	7
4	10
5	13
6	16
50	148
n	3n-2

Building Number	Number of Blocks Needed
1	1
2	6
3	11
4	16
5	21
6	26
50	246
n	5n-4

When the first set of differences is constant, this constant becomes the coefficient of the general solution.

FOREVER AND EVER, AMEN!
(pg.22)

Number of Terms	Sum
1	1
2	$1\frac{1}{2}$
3	$1\frac{3}{4}$
4	$1\frac{7}{8}$
5	$1\frac{15}{16}$
6	$1\frac{31}{32}$
7	$1\frac{63}{64}$
20	$1\frac{524287}{524288}$

The sum will always be less than 2.
The numbers in the second column will always be 1 and something.

HINGED!
(pg.23)

Number of Polygons	Perimeter of Triangles	Perimeter of Squares	Perimeter of Pentagons	Perimeter of Hexagons
1	3	4	5	6
2	4	6	8	10
3	5	8	11	14
4	6	10	14	18
5	7	12	17	22
10	12	22	32	42
n	n+2	2n+2	3n+2	4n+2

Extra Challenge Solution: $(s - 2)n + 2$
From the table it can be seen that the coefficient of n is always two less than the number of sides of the polygon.

THE PERFECT SHUFFLE
(pg.24)

Number of Cards	Number of Shuffles
4	2
6	4
8	3
10	6

THE SHUFFLE FORMULA
(pg.25)

Students should do "The Perfect Shuffle" activity before this one.

Number of Cards	Number of Shuffles
4	2
6	4
8	3
10	6
12	10
14	12
16	4
22	6
32	5
52	8

SIDESWIPED SQUARES
(pg.26)

Length of One Side	Area of Square	Number of Unit Squares Touching On		
		2 Sides	1 Side	No Sides
2	4	4	0	0
3	9	4	4	1
4	16	4	8	4
5	25	4	12	9
6	36	4	16	16
7	49	4	20	25
10	100	4	32	64
n	n^2	4	4n-8 or 4(n-2)	$(n-2)^2$

HARE CITY POPULATION EXPLODES
(pg.27)

Month	Number of Pairs
Jan	1
Feb	1
Mar	2
Apr	3
May	5
Jun	8
Jul	13
Aug	21
Sep	34
Oct	55
Nov	89
Dec	144

The pattern in this problem is the famous Fibonacci sequence where each term is obtained by adding the previous two terms.

CHAIN LETTER MADNESS
(pg.28)

Cycle Number	Number of People Needed for Cycle	Total Number of People Needed from Beginning
1	20	20
2	400	420
3	8000	8420
4	160,000	168,420
5	3,200,00	3,368,420
6	64,000,00	67,368,420
7	1,280,000,000	1,347,368,420
8	25,600,000,000	26,947,368,420

This activity is particularly interesting to students because many of them have already had experience with chain letters. Teachers may wish to point out that the number of people needed to complete a mere 8 cycles of the letter is more than five times the current population of the world!

PATTERNS WITH A POINT
(pg.29)

a. 4
b. 1
c. 8 (Hint: Because there are three digits in the repeating pattern, divide 50 by 3. The remainder of 2 points to the appropriate digit.)

d.

Decimal Place	Digit
1	2
2	3
3	2
4	3
5	2
6	3
50	3

e.

Decimal Place	Digit
1	1
2	7
3	1
4	7
5	1
6	7
50	7

f.

Decimal Place	Digit
1	2
2	9
3	6
4	2
5	9
6	6
50	9

g.

Decimal Place	Digit
1	2
2	7
3	2
4	7
5	2
6	7
50	7

MAGIC MULTIPLICATION
(pg.30)

The last two digits in each product is the product of the units digits of each number. Everything but the last two digits is obtained by multiplying the tens digit by a number one larger.

PIERCED POLYGONS
(pg.31)

Number of Triangles = m + n

Figure	m	n	Number of Triangles
1	4	3	7
2	5	4	9
3	3	3	6
4	4	4	8
5	7	5	12
6	6	4	10

SUM WILL, SUM WON'T
(pg.32)

Number of Dice	Smallest Possible Sum	Largest Possible Sum	Sum(s) With Greatest Chance of Occurring
2	2	12	7
3	3	18	10 or 11
4	4	24	14
5	5	30	17 or 18
6	6	36	21
7	7	42	24 or 25
8	8	48	28
9	9	54	31 or 32
20	20	120	70
40	40	240	140
100	100	600	350
n (even)	n	6n	$\frac{7n}{2}$
n (odd)	n	6n	$\frac{7n-1}{2}$ or $\frac{7n+1}{2}$

THE END OF THE WORLD
(pg.33)

Number of Disks	Minimum Number of Moves
1	1
2	3
3	7
4	15
5	31
64	$2^{64} - 1$
n	$2^n - 1$

Extra Challenge Solution:
Approximately $5.85 \times 10^{11} = 585{,}000{,}000{,}000$ years. Students will need a scientific calculator to solve the extra challenge.

CIRCLE CIRCUS
(pg.34)

Number of Circles	Maximum Number of Regions
1	1
2	3
3	7
4	13
5	21
6	31
7	43
8	57
20	301
n	$n^2 - n + 1$

The first set of differences for the numbers in the second column is 2, 4, 6, 8, 10, etc. This pattern allows students to continue the table indefinitely. The **horizontal** solution is $n(n-1) + 1 = n^2 - n + 1$. To help students, suggest they subtract one from each number in the second column, obtaining 0, 2, 6, 12, 20, etc.

TWENTY-ONE CONNECT
(pg.35)

Number of Points	Number of Segments
2	1
3	3
4	6
5	10
6	15
7	21
21	210
n	$\frac{n(n-1)}{2}$

Students should count the number of segments for circles with 5 points and 6 points. To be sure all possible segments are drawn, have students examine how many originate from each point. This should always be one fewer than the number of points on the circle.

Note that the differences for the numbers in the second column are 1, 2, 3, 4, 5, etc. This discovery allows students to continue the table indefinitely. If students are familiar with triangular numbers, they will recognize the numbers in the second column as triangular numbers.

HEXAGONAL NUMBERS
(pg.36)

Some students may discover that the numbers in the second column are every other triangular number—**1**, 3, **6**, 10, **15**, 21, **28**, 36, **45**, . . .
Also, the first set of differences for the numbers in the second column is 5, 9, 13, 17, 21, etc., increasing by 4 each time. This discovery allows them to continue the numbers in the second column indefinitely.

Hexagonal Number	Number of Dots
1st	1
2nd	6
3rd	15
4th	28
5th	45
6th	66
7th	91
8th	120

Extra Challenge Solution: $2n^2 - n$

Suggest that students divide the number in the second column by the corresponding number in the first column, obtaining 1, 3, 5, 7, 9, 11, etc. It is now quite easy to obtain the general solution $2n^2 - n$ as shown in the following table.

Hexagonal Number	Number of Dots	2nd Column Number ÷ 1st Column Number
1st	1	1
2nd	6	3
3rd	15	5
4th	28	7
5th	45	9
6th	66	11
nth	n(2n-1)	2n-1

If students have some algebra background the method of finite differences (see *Appendix)* may also be used to find the general solution.

FRIGHTENING FRACTIONS
(pg.37)

Number of Terms	Sum
1	$\frac{1}{2}$
2	$\frac{2}{3}$
3	$\frac{3}{4}$
4	$\frac{4}{5}$
5	$\frac{5}{6}$
6	$\frac{6}{7}$
7	$\frac{7}{8}$
50	$\frac{50}{51}$

Extra Challenge Solution: $\frac{n}{n+1}$

MORE TO SEE THAN ONE, TWO, THREE!
(pg.38)

Row Number	Sum
1	1
2	4
3	9
4	16
5	25
6	36
7	49
n	n^2

Number of Rows in Array	Sum of all Numbers in Array	Helping Column
1	1	1•2•3
2	5	2•3•5
3	14	3•4•7
4	30	4•5•9
5	55	5•6•11
6	91	6•7•13
7	140	7•8•15
n	$\frac{n(n+1)(2n+1)}{6}$	n(n+1) (2n+1)

ONE POTATO, TWO...
(pg.39)

The first set of differences for the numbers in the second column is 1, 2, 4, 7, . . . The second set of differences is 1, 2, 3, . . . Using this pattern, the table can be extended indefinitely.

Number of Cuts	Maximum Number of Pieces
0	1
1	2
2	4
3	8
4	15
5	26
6	42
7	64
8	93

Extra Challenge Solution: $\frac{n^3 + 5n + 6}{6}$

The horizontal solution for this problem is quite challenging. If students have some algebra background it may be obtained using the method of finite differences (see *Appendix).*

Another approach is to give students some help by asking them to examine the following:

$0(0^2 + 5) + 6 = 6$
$1(1^2 + 5) + 6 = 12$
$2(2^2 + 5) + 6 = 24$
$3(3^2 + 5) + 6 = 48$
$4(4^2 + 5) + 6 = 90$
etc.

From this, students should obtain $n(n^2 + 5) + 6 = n^3 + 5n + 6$. Since the result is always 6 times the numbers in the second column of the table, the solution is $\frac{(n^3 + 5n + 6)}{6}$.

PASCAL WINS THE WORLD SERIES
(pg.40)

```
                        1
                      1   1
                    1   2  (1)
                  1   3  (3)  1
                1   4   6  (4)  1
              1   5  10 (10)  5   1
            1   6  15  20 (15)  6   1
          1   7  21  35 (35) 21   7   1
        1   8  28  56  70 (56) 28   8   1
```

The pattern to this problem is a zig-zag path in Pascal's triangle as shown.

Best out of	Number of Possibilities
2	1
3	3
4	4
5	10
6	15
7	35

SECRETS WORTH SHARING
(pg.41)

1. 11 • 11 = 121
   111 • 111 = 12321
   1111 • 1111 = 1 2 3 4 3 2 1
   11111 • 11111 = 1 2 3 4 5 4 3 2 1
   111111 • 111111 = 1 2 3 4 5 6 5 4 3 2 1

2. 1 • 8 + 1 = 9
   12 • 8 + 2 = 98
   123 • 8 + 3 = 9 8 7
   1234 • 8 + 4 = 9 8 7 6
   12345 • 8 + 5 = 9 8 7 6 5
   123456 • 8 + 6 = 9 8 7 6 5 4
   1234567 • 8 + 7 = 9 8 7 6 5 4 3
   12345678 • 8 + 8 = 9 8 7 6 5 4 3 2
   123456789 • 8 + 9 = 9 8 7 6 5 4 3 2 1

3. 1 • 9 + 2 = 11
   12 • 9 + 3 = 111
   123 • 9 + 4 = 1111
   1234 • 9 + 5 = 11111
   12345 • 9 + 6 = 111111
   123456 • 9 + 7 = 1111111

4. 9 • 9 = 81
   99 • 99 = 9801
   999 • 999 = 9 9 8 0 0 1
   9999 • 9999 = 9 9 9 8 0 0 0 1
   99999 • 99999 = 9 9 9 9 8 0 0 0 0 1

THIS LINE'S BUSY!
(pg.42)

Number of Interior Points	Total Number of Segments
0	1
1	3
2	6
3	10
4	15
5	21
6	28
7	36

Extra Challenge
Solution: $\frac{(n+1)(n+2)}{2}$

THE TWELVE DAYS OF CHRISTMAS
(pg.43)

Day	Total Number of Gifts Given
1st	1
2nd	3
3rd	6
4th	10
5th	15
6th	21
7th	28
8th	36
9th	45
10th	55
11th	66
12th	78

Extra Challenge
Solution: $\frac{n(n+1)}{2}$

SAVE THE BEST FOR LAST
(pg.44)

$9^1 = 9$
$9^2 = 81$
$9^3 = 729$
$9^4 = 6561$
$9^5 = 59049$
$9^6 = 531441$

Exponent of 9	Last Digit
1	9
2	1
3	9
4	1
5	9
6	1
12314	1

Exponent of 4	Last Digit
1	4
2	6
3	4
4	6
1289	4

Exponent of 6	Last Digit
1	6
2	6
3	6
4	6
5633	6

Exponent of 5	Last Digit
1	5
2	5
3	5
4	5
823	5

The AIMS Program

AIMS is the acronym for "**A**ctivities **I**ntegrating **M**athematics and **S**cience." Such integration enriches learning and makes it meaningful and holistic. AIMS began as a project of Fresno Pacific University to integrate the study of mathematics and science in grades K-9, but has since expanded to include language arts, social studies, and other disciplines.

AIMS is a continuing program of the non-profit AIMS Education Foundation. It had its inception in a National Science Foundation funded program whose purpose was to explore the effectiveness of integrating mathematics and science. The project directors in cooperation with 80 elementary classroom teachers devoted two years to a thorough field-testing of the results and implications of integration.

The approach met with such positive results that the decision was made to launch a program to create instructional materials incorporating this concept. Despite the fact that thoughtful educators have long recommended an integrative approach, very little appropriate material was available in 1981 when the project began. A series of writing projects have ensued, and today the AIMS Education Foundation is committed to continue the creation of new integrated activities on a permanent basis.

The AIMS program is funded through the sale of books, products, and staff development workshops and through proceeds from the Foundation's endowment. All net income from program and products flows into a trust fund administered by the AIMS Education Foundation. Use of these funds is restricted to support of research, development, and publication of new materials. Writers donate all their rights to the Foundation to support its on-going program. No royalties are paid to the writers.

The rationale for integration lies in the fact that science, mathematics, language arts, social studies, etc., are integrally interwoven in the real world from which it follows that they should be similarly treated in the classroom where we are preparing students to live in that world. Teachers who use the AIMS program give enthusiastic endorsement to the effectiveness of this approach.

Science encompasses the art of questioning, investigating, hypothesizing, discovering, and communicating. Mathematics is the language that provides clarity, objectivity, and understanding. The language arts provide us powerful tools of communication. Many of the major contemporary societal issues stem from advancements in science and must be studied in the context of the social sciences. Therefore, it is timely that all of us take seriously a more holistic mode of educating our students. This goal motivates all who are associated with the AIMS Program. We invite you to join us in this effort.

Meaningful integration of knowledge is a major recommendation coming from the nation's professional science and mathematics associations. The American Association for the Advancement of Science in *Science for All Americans* strongly recommends the integration of mathematics, science, and technology. The National Council of Teachers of Mathematics places strong emphasis on applications of mathematics such as are found in science investigations. AIMS is fully aligned with these recommendations.

Extensive field testing of AIMS investigations confirms these beneficial results:

1. Mathematics becomes more meaningful, hence more useful, when it is applied to situations that interest students.
2. The extent to which science is studied and understood is increased, with a significant economy of time, when mathematics and science are integrated.
3. There is improved quality of learning and retention, supporting the thesis that learning that is meaningful and relevant is more effective.
4. Motivation and involvement are increased dramatically as students investigate real-world situations and participate actively in the process.

We invite you to become part of this classroom teacher movement by using an integrated approach to learning and sharing any suggestions you may have. The AIMS Program welcomes you!

AIMS Education Foundation Programs

Practical proven strategies to improve student achievement

When you host an AIMS workshop for elementary and middle school educators, you will know your teachers are receiving effective usable training they can apply in their classrooms immediately.

Designed for teachers—AIMS Workshops:

- Correlate to your state standards;
- Address key topic areas, including math content, science content, problem solving, and process skills;
- Teach you how to use AIMS' effective hands-on approach;
- Provide practice of activity-based teaching;
- Address classroom management issues, higher-order thinking skills, and materials;
- Give you AIMS resources; and
- Offer college (graduate-level) credits for many courses.

Aligned to district and administrator needs—AIMS workshops offer:

- Flexible scheduling and grade span options;
- Custom (one-, two-, or three-day) workshops to meet specific schedule, topic and grade-span needs;
- Pre-packaged one-day workshops on most major topics—only $3900 for up to 30 participants (includes all materials and expenses);
- Prepackaged four- or five-day workshops for in-depth math and science training—only $12,300 for up to 30 participants (includes all materials and expenses);
- Sustained staff development, by scheduling workshops throughout the school year and including follow-up and assessment;
- Eligibility for funding under the Title I and Title II sections of No Child Left Behind; and
- Affordable professional development—save when you schedule consecutive-day workshops.

University Credit—Correspondence Courses

AIMS offers correspondence courses through a partnership with Fresno Pacific University.

- Convenient distance-learning courses—you study at your own pace and schedule. No computer or Internet access required!

The tuition for each three-semester unit graduate-level course is $264 plus a materials fee.

The AIMS Instructional Leadership Program

This is an AIMS staff-development program seeking to prepare facilitators for leadership roles in science/math education in their home districts or regions. Upon successful completion of the program, trained facilitators become members of the AIMS Instructional Leadership Network, qualified to conduct AIMS workshops, teach AIMS in-service courses for college credit, and serve as AIMS consultants. Intensive training is provided in mathematics, science, process and thinking skills, workshop management, and other relevant topics.

Introducing AIMS Science Core Curriculum

Developed to meet 100% of your state's standards, AIMS' Science Core Curriculum gives students the opportunity to build content knowledge, thinking skills, and fundamental science processes.

- *Each* grade specific module has been developed to extend the AIMS approach to full-year science programs.
- *Each* standards-based module includes math, reading, hands-on investigations, and assessments.

Like all AIMS resources, these core modules are able to serve students at all stages of readiness, making these a great value across the grades served in your school.

For current information regarding the programs described above, please complete the following form and mail it to: P.O. Box 8120, Fresno, CA 93747.

Information Request

Please send current information on the items checked:

_____ *Basic Information Packet* on AIMS materials
_____ *AIMS Instructional Leadership Program*
____ Hosting information for AIMS workshops
____ AIMS Science Core Curriculum

Name ______________________ Phone ______________________

Address ____
Street City State Zip

AIMS Program Publications

Actions with Fractions, 4-9
Awesome Addition and Super Subtraction, 2-3
Bats Incredible! 2-4
Brick Layers II, 4-9
Chemistry Matters, 4-7
Counting on Coins, K-2
Cycles of Knowing and Growing, 1-3
Crazy about Cotton, 3-7
Critters, 2-5
Electrical Connections, 4-9
Exploring Environments, K-6
Fabulous Fractions, 3-6
Fall into Math and Science, K-1
Field Detectives, 3-6
Finding Your Bearings, 4-9
Floaters and Sinkers, 5-9
From Head to Toe, 5-9
Fun with Foods, 5-9
Glide into Winter with Math and Science, K-1
Gravity Rules! 5-12
Hardhatting in a Geo-World, 3-5
It's About Time, K-2
It Must Be A Bird, Pre-K-2
Jaw Breakers and Heart Thumpers, 3-5
Looking at Geometry, 6-9
Looking at Lines, 6-9
Machine Shop, 5-9
Magnificent Microworld Adventures, 5-9
Marvelous Multiplication and Dazzling Division, 4-5
Math + Science, A Solution, 5-9
Mostly Magnets, 2-8
Movie Math Mania, 6-9
Multiplication the Algebra Way, 6-8
Off the Wall Science, 3-9
Out of This World, 4-8
Paper Square Geometry:
The Mathematics of Origami, 5-12
Puzzle Play, 4-8
Pieces and Patterns, 5-9
Popping With Power, 3-5
Positive vs. Negative, 6-9
Primarily Bears, K-6
Primarily Earth, K-3
Primarily Physics, K-3
Primarily Plants, K-3
Problem Solving: Just for the Fun of It! 4-9
Problem Solving: Just for the Fun of It! Book Two, 4-9
Proportional Reasoning, 6-9
Ray's Reflections, 4-8
Sense-Able Science, K-1
Soap Films and Bubbles, 4-9
Solve It! K-1: Problem-Solving Strategies, K-1
Solve It! 2nd: Problem-Solving Strategies, 2
Solve It! 3rd: Problem-Solving Strategies, 3
Solve It! 4th: Problem-Solving Strategies, 4
Solve It! 5th: Problem-Solving Strategies, 5
Spatial Visualization, 4-9
Spills and Ripples, 5-12
Spring into Math and Science, K-1
The Amazing Circle, 4-9
The Budding Botanist, 3-6
The Sky's the Limit, 5-9
Through the Eyes of the Explorers, 5-9
Under Construction, K-2
Water Precious Water, 2-6
Weather Sense: Temperature, Air Pressure, and Wind, 4-5
Weather Sense: Moisture, 4-5
Winter Wonders, K-2

Spanish Supplements*
Fall Into Math and Science, K-1
Glide Into Winter with Math and Science, K-1
Mostly Magnets, 2-8
Pieces and Patterns, 5-9
Primarily Bears, K-6
Primarily Physics, K-3
Sense-Able Science, K-1
Spring Into Math and Science, K-1

* Spanish supplements are only available as downloads from the AIMS website. The supplements contain only the student pages in Spanish; you will need the English version of the book for the teacher's text.

Spanish Edition
Constructores II: Ingeniería Creativa Con Construcciones LEGO® 4-9
The entire book is written in Spanish. English pages not included.

Other Publications
Historical Connections in Mathematics, Vol. I, 5-9
Historical Connections in Mathematics, Vol. II, 5-9
Historical Connections in Mathematics, Vol. III, 5-9
Mathematicians are People, Too
Mathematicians are People, Too, Vol. II
What's Next, Volume 1, 4-12
What's Next, Volume 2, 4-12
What's Next, Volume 3, 4-12

For further information write to:
AIMS Education Foundation • P.O. Box 8120 • Fresno, California 93747-8120
www.aimsedu.org • 559.255.6396 (fax) • 888.733.2467 (toll free)

Duplication Rights

Standard Duplication Rights

Purchasers of AIMS activities (individually or in books and magazines) may make up to 200 copies of any portion of the purchased activities, provided these copies will be used for educational purposes and only at one school site.

Workshop or conference presenters may make one copy of a purchased activity for each participant, with a limit of five activities per workshop or conference session.

Standard duplication rights apply to activities received at workshops, free sample activities provided by AIMS, and activities received by conference participants.

All copies must bear the AIMS Education Foundation copyright information.

Unlimited Duplication Rights

To ensure compliance with copyright regulations, AIMS users may upgrade from standard to unlimited duplication rights. Such rights permit unlimited duplication of purchased activities (including revisions) for use at a given school site.

Activities received at workshops are eligible for upgrade from standard to unlimited duplication rights.

Free sample activities and activities received as a conference participant are not eligible for upgrade from standard to unlimited duplication rights.

Upgrade Fees

The fees for upgrading from standard to unlimited duplication rights are:
- $5 per activity per site,
- $25 per book per site, and
- $10 per magazine issue per site.

The cost of upgrading is shown in the following examples:
- activity: 5 activities x 5 sites x $5 = $125
- book: 10 books x 5 sites x $25 = $1250
- magazine issue: 1 issue x 5 sites x $10 = $50

Purchasing Unlimited Duplication Rights

To purchase unlimited duplication rights, please provide us the following:
1. The name of the individual responsible for coordinating the purchase of duplication rights.
2. The title of each book, activity, and magazine issue to be covered.
3. The number of school sites and name of each site for which rights are being purchased.
4. Payment (check, purchase order, credit card)

Requested duplication rights are automatically authorized with payment. The individual responsible for coordinating the purchase of duplication rights will be sent a certificate verifying the purchase.

Internet Use

Permission to make AIMS activities available on the Internet is determined on a case-by-case basis.

• P. O. Box 8120, Fresno, CA 93747-8120 •
• permissions@aimsedu.org • www.aimsedu.org •
• 559.255.6396 (fax) • 888.733.2467 (toll free) •